D1484568

Complete Electronics

Complete Electronics

SELF-TEACHING GUIDE WITH PROJECTS

Earl Boysen | **Harry Kybett**

WILEY

John Wiley & Sons, Inc.

Complete Electronics

EXECUTIVE EDITOR: CAROL LONG
PROJECT EDITOR: KEVIN SHAFER
TECHNICAL EDITOR: REX MILLER
PRODUCTION EDITOR: KATHLEEN WISOR
COPY EDITOR: SAN DEE PHILLIPS
EDITORIAL MANAGER: MARY BETH WAKEFIELD
FREELANCER EDITORIAL MANAGER: ROSEMARIE GRAHAM
ASSOCIATE DIRECTOR OF MARKETING: DAVID MAYHEW
MARKETING MANAGER: ASHLEY ZURCHER
BUSINESS MANAGER: AMY KNIES
PRODUCTION MANAGER: TIM TATE
VICE PRESIDENT AND EXECUTIVE GROUP PUBLISHER: RICHARD SWADLEY
VICE PRESIDENT AND EXECUTIVE PUBLISHER: NEIL EDDE
ASSOCIATE PUBLISHER: JIM MINATEL
PROJECT COORDINATOR, COVER: KATIE CROCKER
PROOFREADER: NANCY CARRASCO
INDEXER: JACK LEWIS
COVER IMAGE: © BART COENDERS / ISTOCKPHOTO
COVER DESIGNER: RYAN SNEED

Published by
John Wiley & Sons, Inc.
10475 Crosspoint Boulevard
Indianapolis, IN 46256
www.wiley.com

Copyright © 2012 by Earl Boysen and Harry Kybett

Published by John Wiley & Sons, Inc., Indianapolis, Indiana

Published simultaneously in Canada

ISBN: 978-1-118-21732-0
ISBN: 978-1-118-28232-8 (ebk)
ISBN: 978-1-118-28319-6 (ebk)
ISBN: 978-1-118-28469-8 (ebk)

Manufactured in the United States of America

10 9 8 7 6 5 4 3

To my science and engineering teachers. I'd particularly like to thank Jim Giovando, my physics and chemistry teacher at Petaluma Senior High School, who, even decades later, I remember as having been an inspiration. I also dedicate this book to the physics and chemistry faculty of Sonoma State University in the 1970s, where the small class size and personal guidance by the professors made for a great learning environment.

About the Author

Earl Boysen spent 20 years as an engineer in the semiconductor industry, and currently runs two websites, BuildingGadgets.com (dedicated to electronics) and UnderstandingNano.com (covering nanotechnology topics). Boysen holds a Masters degree in Engineering Physics from the University of Virginia. He is the co-author of three other books: *Electronics Projects For Dummies* (Indianapolis: Wiley, 2006), *Nanotechnology For Dummies* (Indianapolis: Wiley, 2011), and the first edition of *Electronics For Dummies* (Indianapolis: Wiley, 2005). He lives with his wonderful wife, Nancy, and two cats.

About the Technical Editor

Rex Miller was a Professor of Industrial Technology at The State University of New York, College at Buffalo for more than 35 years. He has taught on the technical school, high school, and college level for more than 40 years. He is the author or co-author of more than 100 textbooks ranging from electronics through carpentry and sheet metal work. He has contributed more than 50 magazine articles over the years to technical publications. He is also the author of seven civil war regimental histories.

Acknowledgments

I want to first thank Harry Kybett for authoring the original version of this book many years ago. It's an honor to take over such a classic book in the electronics field. Thanks also to Carol Long for bringing me on board with the project, and Kevin Shafer for his able project management of the book. My appreciation to Rex Miller for his excellent technical editing, and to San Dee Phillips for handling all the mechanics of spelling and grammar in a thorough copy edit. Finally, thanks to my wonderful wife, Nancy, for her advice and support throughout the writing of this book.

Contents at a Glance

Introduction

The rapid growth of modern electronics is truly a phenomenon. Electronic devices (including cell phones, personal computers, portable MP3 players, and digital cameras) are a big part of many of our daily lives. Many industries have been founded, and older industries have been revamped, because of the availability and application of modern electronics in manufacturing processes, as well as in electronic products. Electronic products are constantly evolving, and their impact on our lives, and even the way we socialize, is substantial.

WHAT THIS BOOK TEACHES

Complete Electronics Self-Teaching Guide with Projects is for anyone who has a basic understanding of electronics concepts and wants to understand the operation of components found in the most common discrete circuits. The chapters focus on circuits that are the building blocks for many common electronic devices, and on the very few (but important) principles you need to know to work with electronics.

The arrangement and approach is completely different from any other book on electronics in that it uses a question-and-answer approach to help you understand how electronic circuits work. This book steps you through calculations for every example in an easy-to-understand fashion, and you do not need to have a mathematical background beyond first-year algebra to follow along.

For many of you, the best way to understand new concepts is by doing, rather than reading or listening. This book reinforces your understanding of electronic concepts by leading you through the calculations and concepts for key circuits, as well as the construction of circuits. Projects interspersed throughout the material enable you to get hands-on practice. You build many of the circuits and observe or measure how they work.

Helpful sidebars are interspersed throughout the book to provide more information about how components work, and how to choose the right component. Other sidebars provide discussions of techniques for building and testing circuits. If you want this additional information, be sure to read these.

Understanding the circuits composed of discrete components and the applicable calculations discussed is useful not only in building and designing circuits, but it also helps you to work with integrated circuits (ICs). That's because ICs use miniaturized components (such as transistors, diodes, capacitors, and resistors) that function based on the same rules as discrete components (along with some specific rules necessitated by the extremely small size of IC components).

HOW THIS BOOK IS ORGANIZED

This book is organized with sets of problems that challenge you to think through a concept or procedure, and then provides answers so that you can constantly check your progress and understanding. Specifically, the chapters in this book are organized as follows:

CHAPTER 1 DC Review and Pre-Test—This chapter provides a review and pre-test on the basic concepts, components, and calculations that are useful when working with direct current (DC) circuits.

CHAPTER 2 The Diode—This chapter teaches you about the diode, including how you use diodes in DC circuits, the main characteristics of diodes, and calculations you can use to determine current, voltage, and power.

CHAPTER 3 Introduction to the Transistor—This chapter explores the transistor and how it's used in circuits. You also discover how bipolar junction transistors (BJTs) and junction field effect transistors (JFETs) control the flow of electric current.

CHAPTER 4 The Transistor Switch—This chapter examines the simplest and most widespread application of the transistor: switching. In addition to learning how to design a transistor circuit to drive a particular load, you also compare the switching action of a JFET and a BJT.

CHAPTER 5 AC Pre-Test and Review—This chapter examines the basic concepts and equations for alternating current (AC) circuits. You discover how to use resistors and capacitors in AC circuits, and learn related calculations.

CHAPTER 6 Filters—This chapter looks at how resistors, capacitors, and inductors are used in high-pass filters and low-pass filters to pass or block AC signals above or below a certain frequency.

CHAPTER 7 Resonant Circuits—This chapter examines the use of capacitors, inductors, and resistors in bandpass filters and band-reject filters to pass or block AC signals in a band of frequencies. You also learn how to calculate the resonance frequency and bandwidth of these circuits. This chapter also introduces the use of resonant circuits in oscillators.

CHAPTER 8 Transistor Amplifiers—This chapter explores the use of transistor amplifiers to amplify electrical signals. In addition to examining the fundamental steps used to design BJT-based amplifiers, you learn how to use JFETs and operational amplifiers (op-amps) in amplifier circuits.

CHAPTER 9 Oscillators —This chapter introduces you to the oscillator, a circuit that produces a continuous AC output signal. You learn how an oscillator works and step through the procedure to design and build an oscillator.

CHAPTER 10 The Transformer—This chapter discusses how a transformer converts AC voltage to a higher or lower voltage. You learn how a transformer makes this conversion and how to calculate the resulting output voltage.

CHAPTER 11 Power Supply Circuits—This chapter examines how power supplies convert AC to DC with a circuit made up of transformers, diodes, capacitors, and resistors. You also learn how to calculate the values of components that produce a specified DC output voltage for a power supply circuit.

CHAPTER 12 Conclusion and Final Self-Test—This chapter enables you to check your overall knowledge of electronics concepts presented in this book through the use of a final self-test.

In addition, this book contains the following appendixes for easy reference:
APPENDIX A Glossary—This appendix provides key electronics terms and their definitions.

APPENDIX B List of Symbols and Abbreviations—This appendix gives you a handy reference of commonly used symbols and abbreviations.

APPENDIX C Powers of Ten and Engineering Prefixes—This appendix lists prefixes commonly used in electronics, along with their corresponding values.

APPENDIX D Standard Resistor Values—This appendix provides standard resistance values for the carbon film resistor, the most commonly used type of resistor.

APPENDIX E Supplemental Resources—This appendix provides references to helpful websites, books, and magazines.

APPENDIX F Equation Reference—This appendix serves as a quick guide to commonly used equations, along with chapter and problem references showing you where they are first introduced in this book.

APPENDIX G Schematic Symbols Used in This Book—This appendix provides a listing of schematic symbols used in the problems found throughout the book.

CONVENTIONS USED IN THIS BOOK

As you study electronics, you will find that there is some variation in terminology and the way that circuits are drawn. Following are three conventions followed in this book that you should be aware of:

- The discussions use "V" to stand for voltage, versus "E," which you see used in some other books.

- In all circuit diagrams, intersecting lines indicate an electrical connection. (Some other books use a dot at the intersection of lines to indicate a connection.) If a semicircle appears at the intersection of two lines, it indicates that there is no connection. See Figure 9.5 for an example of this.

- The discussions in this book use conventional current flow to determine the flow of electric current (from positive voltage to negative voltage), whereas some other books use electron flow (from negative voltage to positive voltage).

HOW TO USE THIS BOOK

This book assumes that you have some knowledge of basic electronics such as Ohm's law and current flow. If you have read a textbook or taken a course on electronics, or if you have worked with electronics, you probably have the prerequisite knowledge. If not, you should read a book such as *Electronics for Dummies* (Indianapolis: Wiley, 2009) to get the necessary background for this book. You can also go to the author's Website (www.BuildingGadgets.com) and use the Tutorial links to find useful online lessons in electronics. In addition, Chapters 1 and 5 enable you to test your knowledge and review the necessary basics of electronics.

You should read the chapters in order because often later material depends on concepts and skills covered in earlier chapters.

Complete Electronics Self-Teaching Guide with Projects is presented in a self-teaching format that enables you to learn easily, and at your own pace. The material is presented in numbered sections called *problems*. Each problem presents some new information and gives you questions to answer. To learn most effectively, you should cover up the answers with a sheet of paper and try to answer each question. Then, compare your answer with the correct answer that follows. If you miss a question, correct your answer and then go on. If you miss many in a row, go back and review the previous section, or you may miss the point of the material that follows.

Be sure to try to do all the projects. They are not difficult, and they help reinforce your learning of the subject matter. If you don't have the equipment to work through a project, simply reading through it can help you to better understand the concepts it demonstrates.

Each project includes a schematic, parts list, step-by-step instructions, and detailed photos of the completed circuit. Working through these projects, you can test your skill by building the circuit using just the schematic and parts list. If you want additional help, check the photos showing the details of how the components are connected. A Camera icon in the margin as shown here indicates that there is a color version of the figure in a special insert in the paperback version of this book. If you purchased an electronic version of this book, and have an e-reader without color capabilities, you can find the color photos on the author's website at www.buildinggadgets.com/complete-electronics.htm.

This website also provides project pages that include links to suppliers. These pages are kept up-to-date with supplier part numbers for the components you need.

When you reach the end of a chapter, evaluate your learning by taking the Self-Test. If you miss any questions, review the related parts of the chapter again. If you do well on the Self-Test, you're ready to go to the next chapter. You may also find the Self-Test useful as a review before you start the next chapter. At the end of the book, there is a Final Self-Test that enables you to assess your overall learning.

You can work through this book alone, or you can use it with a course. If you use the book alone, it serves as an introduction to electronics but is not a complete course. For that reason, at the end of the book are some suggestions for further reading and online resources. Also, at the back of the book is a table of symbols and abbreviations for reference and review.

Now you're ready to learn *electronics*!

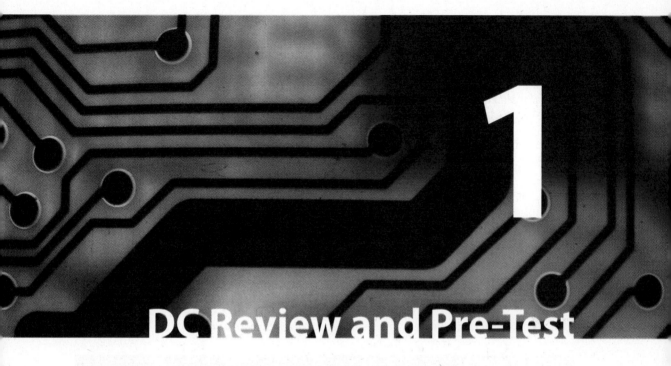

1

DC Review and Pre-Test

Electronics cannot be studied without first understanding the basics of electricity. This chapter is a review and pre-test on those aspects of direct current (DC) that apply to electronics. By no means does it cover the whole DC theory, but merely those topics that are essential to simple electronics.

This chapter reviews the following:

- Current flow
- Potential or voltage difference
- Ohm's law
- Resistors in series and parallel

- Power
- Small currents
- Resistance graphs
- Kirchhoff's Voltage Law
- Kirchhoff's Current Law
- Voltage and current dividers
- Switches
- Capacitor charging and discharging
- Capacitors in series and parallel

CURRENT FLOW

1 Electrical and electronic devices work because of an electric current.

QUESTION

What is an electric current? _____

_____

ANSWER

An *electric current* is a flow of electric charge. The electric charge usually consists of negatively charged electrons. However, in semiconductors, there are also positive charge carriers called *holes*.

2 There are several methods that can be used to generate an electric current.

QUESTION

Write at least three ways an electron flow (or current) can be generated. _____

_____

_____

_____

ANSWER

The following is a list of the most common ways to generate current:

- *Magnetically*—This includes the induction of electrons in a wire rotating within a magnetic field. An example of this would be generators turned by water, wind, or steam, or the fan belt in a car.

- *Chemically*—This involves the electrochemical generation of electrons by reactions between chemicals and electrodes (as in batteries).

- *Photovoltaic generation of electrons*—This occurs when light strikes semiconductor crystals (as in solar cells).

Less common methods to generate an electric current include the following:

- *Thermal generation*—This uses temperature differences between thermocouple junctions. Thermal generation is used in generators on spacecrafts that are fueled by radioactive material.

- *Electrochemical reaction*—This occurs between hydrogen, oxygen, and electrodes (fuel cells).

- *Piezoelectrical*—This involves mechanical deformation of piezoelectric substances. For example, piezoelectric material in the heels of shoes power LEDs that light up when you walk.

3 Most of the simple examples in this book contain a battery as the voltage source. As such, the source provides a potential difference to a circuit that enables a current to flow. An *electric current* is a flow of electric charge. In the case of a battery, electrons are the electric charge, and they flow from the terminal that has an excess number of electrons to the terminal that has a deficiency of electrons. This flow takes place in any complete circuit that is connected to battery terminals. It is this difference in the charge that creates the potential difference in the battery. The electrons try to balance the difference.

Because electrons have a negative charge, they actually flow from the negative terminal and return to the positive terminal. This direction of flow is called *electron flow*. Most books, however, use current flow, which is in the opposite direction. It is referred to as *conventional current flow*, or simply current flow. In this book, the term conventional current flow is used in all circuits.

Later in this book, you see that many semiconductor devices have a symbol that contains an arrowhead pointing in the direction of conventional current flow.

QUESTIONS

A. Draw arrows to show the current flow in Figure 1.1. The symbol for the battery shows its polarity.

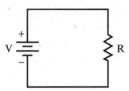

FIGURE 1.1

B. What indicates that a potential difference is present? _____
_____

C. What does the potential difference cause? _____
_____

D. What will happen if the battery is reversed? _____
_____

ANSWERS

A. See Figure 1.2.

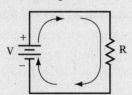

FIGURE 1.2

B. The battery symbol indicates that a difference of potential (also called *voltage*) is being supplied to the circuit.

C. Voltage causes current to flow if there is a complete circuit present, as shown in Figure 1.1.

D. The current flows in the opposite direction.

OHM'S LAW

4 Ohm's law states the fundamental relationship between voltage, current, and resistance.

QUESTION

What is the algebraic formula for Ohm's law? _____

_____

ANSWER

$V = I \times R$

This is the most basic equation in electricity, and you should know it well. Some electronics books state Ohm's law as $E = IR$. E and V are both symbols for voltage. This book uses V to indicate voltage. When V is used after a number in equations and circuit diagrams, it represents volts, the unit of measurement of voltage. Also, in this formula, resistance is the opposition to current flow. Larger resistance results in smaller current for any given voltage.

5 Use Ohm's law to find the answers in this problem.

QUESTIONS

What is the voltage for each combination of resistance and current values?

A. R = 20 ohms, I = 0.5 amperes

  V = _____

B. R = 560 ohms, I = 0.02 amperes

  V = _____

C. R = 1,000 ohms, I = 0.01 amperes

  V = _____

D. R = 20 ohms   I = 1.5 amperes

  V = _____

ANSWERS

A. 10 volts

B. 11.2 volts

C. 10 volts

D. 30 volts

6 You can rearrange Ohm's law to calculate current values.

QUESTIONS

What is the current for each combination of voltage and resistance values?

A. V = 1 volt, R = 2 ohms

   I = _____

B. V = 2 volts, R = 10 ohms

   I = _____

C. V = 10 volts, R = 3 ohms

   I = _____

D. V = 120 volts, R = 100 ohms

   I = _____

ANSWERS

A. 0.5 amperes

B. 0.2 amperes

C. 3.3 amperes

D. 1.2 amperes

7 You can rearrange Ohm's law to calculate resistance values.

QUESTIONS

What is the resistance for each combination of voltage and current values?

A. V = 1 volt, I = 1 ampere

R = _____

B. V = 2 volts, I = 0.5 ampere

R = _____

C. V = 10 volts, I = 3 amperes

R = _____

D. V = 50 volts, I = 20 amperes

R = _____

ANSWERS

A. 1 ohm

B. 4 ohms

C. 3.3 ohms

D. 2.5 ohms

8 Work through these examples. In each case, two factors are given and you must find the third.

QUESTIONS

What are the missing values?

A. 12 volts and 10 ohms. Find the current. _____

_____

B. 24 volts and 8 amperes. Find the resistance. _____

_____

C. 5 amperes and 75 ohms. Find the voltage. _____

_____

ANSWERS

A. 1.2 amperes

B. 3 ohms

C. 375 volts

INSIDE THE RESISTOR

Resistors are used to control the current that flows through a portion of a circuit. You can use Ohm's law to select the value of a resistor that gives you the correct current in a circuit. For a given voltage, the current flowing through a circuit increases when using smaller resistor values and decreases when using larger resistor values.

This resistor value works something like pipes that run water through a plumbing system. For example, to deliver the large water flow required by your water heater, you might use a 1-inch diameter pipe. To connect a bathroom sink to the water supply requires much smaller water flow and, therefore, works with a 1/2-inch pipe. In the same way, smaller resistor values (meaning less resistance) increase current flow, whereas larger resistor values (meaning more resistance) decrease the flow.

Tolerance refers to how precise a stated resistor value is. When you buy *fixed resistors* (in contrast to *variable resistors* that are used in some of the projects in this book), they have a particular resistance value. Their tolerance tells you how close to that value their resistance will be. For example, a 1,000-ohm resistor with ± 5 percent tolerance could have a value of anywhere from 950 ohms to 1,050 ohms. A 1,000-ohm resistor with ± 1 percent tolerance (referred to as a *precision resistor*) could have a value ranging anywhere from 990 ohms to 1,010 ohms. Although you are assured that the resistance of a precision resistor will be close to its stated value, the resistor with ± 1 percent tolerance costs more to manufacture and, therefore, costs you more than twice as much as a resistor with ± 5 percent.

Most electronic circuits are designed to work with resistors with ± 5 percent tolerance. The most commonly used type of resistor with ± 5 percent tolerance is called a *carbon film resistor*. This term refers to the manufacturing process in which a carbon film is deposited on an insulator. The thickness and width of the carbon film determines the resistance (the thicker the carbon film, the lower the resistance). Carbon film resistors work well in all the projects in this book.

On the other hand, precision resistors contain a metal film deposited on an insulator. The thickness and width of the metal film determines the resistance. These resistors are called *metal film resistors* and are used in circuits for precision devices such as test instruments.

Resistors are marked with four or five color bands to show the value and tolerance of the resistor, as illustrated in the following figure. The four-band color code is used for most resistors. As shown in the figure, by adding a fifth band, you get a five-band color code. Five-band color codes are used to provide more precise values in precision resistors.

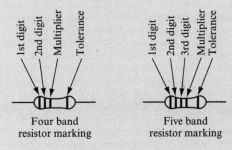

Four band
resistor marking

Five band
resistor marking

The following table shows the value of each color used in the bands:

Color	Significant Digits	Multiplier	Tolerance
Black	0	1	
Brown	1	10	± 1 percent
Red	2	100	± 2 percent
Orange	3	1,000	
Yellow	4	10,000	
Green	5	100,000	
Blue	6	1,000,000	
Violet	7		
Gray	8		
White	9		
Gold		0.1	± 5 percent
Silver		0.01	± 10 percent

Continued

(continued)

> By studying this table, you can see how this code works. For example, if a resistor is marked with orange, blue, brown, and gold bands, its nominal resistance value is 360 ohms with a tolerance of ± 5 percent. If a resistor is marked with red, orange, violet, black, and brown, its nominal resistance value is 237 ohms with a tolerance of ± 1 percent.

RESISTORS IN SERIES

9 You can connect resistors in series. Figure 1.3 shows two resistors in series.

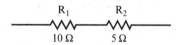

R_1 R_2
$10\ \Omega$ $5\ \Omega$

FIGURE 1.3

QUESTION

What is their total resistance? _____

_____

ANSWER

$R_T = R_1 + R_2 = 10\ \text{ohms} + 5\ \text{ohms} = 15\ \text{ohms}$

The total resistance is often called the *equivalent series resistance* and is denoted as R_{eq}.

RESISTORS IN PARALLEL

10 You can connect resistors in parallel, as shown in Figure 1.4.

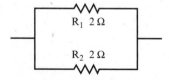

$R_1\ 2\ \Omega$

$R_2\ 2\ \Omega$

FIGURE 1.4

QUESTION

What is the total resistance here? _____

_____

ANSWER

$$\frac{1}{R_T} = \frac{1}{R_1} + \frac{1}{R_2} = \frac{1}{2} + \frac{1}{2} = 1; \text{ thus } R_T = 1 \text{ ohm}$$

R_T is often called the *equivalent parallel resistance*.

11 The simple formula from problem 10 can be extended to include as many resistors as wanted.

QUESTION

What is the formula for three resistors in parallel? _____

_____

ANSWER

$$\frac{1}{R_T} = \frac{1}{R_1} + \frac{1}{R_2} + \frac{1}{R_3}$$

You often see this formula in the following form:

$$R_T = \frac{1}{\dfrac{1}{R_1} + \dfrac{1}{R_2} + \dfrac{1}{R_3}}$$

12 In the following exercises, two resistors are connected in parallel.

QUESTIONS

What is the total or equivalent resistance?

A. $R_1 = 1$ ohm, $R_2 = 1$ ohm

$R_T = $ _____

B. $R_1 = 1,000$ ohms, $R_2 = 500$ ohms

$R_T = $ _____

C. $R_1 = 3,600$ ohms, $R_2 = 1,800$ ohms

$R_T = $ _____

ANSWER

A. 0.5 ohms

B. 333 ohms

C. 1,200 ohms

R_T is always smaller than the smallest of the resistors in parallel.

POWER

13 When current flows through a resistor, it dissipates power, usually in the form of heat. Power is expressed in terms of watts.

QUESTION

What is the formula for power? _____

_____

ANSWER

There are three formulas for calculating power:

$$P = VI \text{ or } P = I^2R \text{ or } P = \frac{V^2}{R}$$

14 The first formula shown in problem 13 allows power to be calculated when only the voltage and current are known.

QUESTIONS

What is the power dissipated by a resistor for the following voltage and current values?

A. V = 10 volts, I = 3 amperes

 P = _____

B. V = 100 volts, I = 5 amperes

 P = _____

C. V = 120 volts, I = 10 amperes

 P = _____

ANSWERS

A. 30 watts.

B. 500 watts, or 0.5 kW. (The abbreviation kW indicates kilowatts.)

C. 1,200 watts, or 1.2 kW.

15 The second formula shown in problem 13 allows power to be calculated when only the current and resistance are known.

QUESTIONS

What is the power dissipated by a resistor given the following resistance and current values?

A. R = 20 ohm, I = 0.5 ampere

 P = _____

B. R = 560 ohms, I = 0.02 ampere

 P = _____

C. V = 1 volt, R = 2 ohms

 P = _____

D. V = 2 volt, R = 10 ohms

 P = _____

ANSWERS

A. 5 watts

B. 0.224 watts

C. 0.5 watts

D. 0.4 watts

16 Resistors used in electronics generally are manufactured in standard values with regard to resistance and power rating. Appendix D shows a table of standard resistance values for 0.25- and 0.05-watt resistors. Quite often, when a certain resistance value is needed in a circuit, you must choose the closest standard value. This is the case in several examples in this book.

You must also choose a resistor with the power rating in mind. Never place a resistor in a circuit that requires that resistor to dissipate more power than its rating specifies.

QUESTIONS

If standard power ratings for carbon film resistors are 1/8, 1/4, 1/2, 1, and 2 watts, what power ratings should be selected for the resistors that were used for the calculations in problem 15?

A. For 5 watts _____

B. For 0.224 watts _____

C. For 0.5 watts _____

D. For 0.4 watts _____

ANSWERS

A. 5 watt (or greater)

B. 1/4 watt (or greater)

C. 1/2 watt (or greater)

D. 1/2 watt (or greater)

Most electronics circuits use low-power carbon film resistors. For higher-power levels (such as the 5-watt requirement in question A), other types of resistors are available.

SMALL CURRENTS

17 Although currents much larger than 1 ampere are used in heavy industrial equipment, in most electronic circuits, only fractions of an ampere are required.

QUESTIONS

A. What is the meaning of the term *milliampere*? _____

B. What does the term *microampere* mean? _____

ANSWERS

A. A milliampere is one-thousandth of an ampere (that is, 1/1000 or 0.001 amperes). It is abbreviated mA.

B. A microampere is one-millionth of an ampere (that is, 1/1,000,000 or 0.000001 amperes). It is abbreviated μA.

18 In electronics, the values of resistance normally encountered are quite high. Often, thousands of ohms and occasionally even millions of ohms are used.

QUESTIONS

A. What does kΩ mean when it refers to a resistor? _____

B. What does MΩ mean when it refers to a resistor? _____

ANSWERS

A. Kilohm (k = kilo, Ω = ohm). The resistance value is thousands of ohms. Thus, 1 kΩ = 1,000 ohms, 2 kΩ = 2,000 ohms, and 5.6 kΩ = 5,600 ohms.

B. Megohm (M = mega, Ω = ohm). The resistance value is millions of ohms. Thus, 1 MΩ = 1,000,000 ohms, and 2.2 MΩ = 2,200,000 ohms.

19 The following exercise is typical of many performed in transistor circuits. In this example, 6 volts is applied across a resistor, and 5 mA of current is required to flow through the resistor.

QUESTIONS

What value of resistance must be used and what power will it dissipate?

R = _____ P = _____

ANSWERS

$$R = \frac{V}{I} = \frac{6 \text{ volts}}{5 \text{ mA}} = \frac{6}{0.005} = 1200 \text{ ohms} = 1.2\,k\Omega$$

$$P = V \times I = 6 \times 0.005 = 0.030 \text{ watts} = 30\,mW$$

20 Now, try these two simple examples.

QUESTIONS

What is the missing value?

A. 50 volts and 10 mA. Find the resistance. _____

_____

B. 1 volt and 1 MΩ. Find the current. _____

_____

ANSWERS

A. 5 kΩ

B. 1 μA

THE GRAPH OF RESISTANCE

21 The voltage drop across a resistor and the current flowing through it can be plotted on a simple graph. This graph is called a *V-I curve*.

Consider a simple circuit in which a battery is connected across a 1 kΩ resistor.

QUESTIONS

A. Find the current flowing if a 10-volt battery is used. _____

_____

B. Find the current when a 1-volt battery is used. _____

C. Now find the current when a 20-volt battery is used. _____

_____

ANSWERS

A. 10 mA

B. 1 mA

C. 20 mA

22 Plot the points of battery voltage and current flow from problem 21 on the graph shown in Figure 1.5, and connect them together.

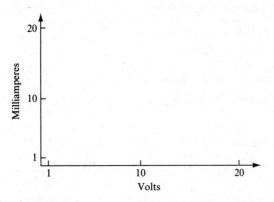

FIGURE 1.5

QUESTION

What would the slope of this line be equal to? _____

_____

ANSWER

You should have drawn a straight line, as in the graph shown in Figure 1.6.

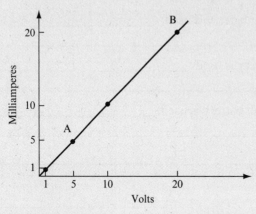

FIGURE 1.6

Sometimes you need to calculate the slope of the line on a graph. To do this, pick two points and call them A and B.

- For point A, let V = 5 volts and I = 5 mA
- For point B, let V = 20 volts and I = 20 mA

The slope can be calculated with the following formula:

$$\text{Slope} = \frac{V_B - V_A}{I_B - I_A} = \frac{20\text{ volts} - 5\text{ volts}}{20\text{ mA} - 5\text{ mA}} = \frac{15\text{ volts}}{15\text{ mA}} = \frac{15\text{ volts}}{0.015\text{ ampere}} = 1\text{ k}\,\Omega$$

In other words, the slope of the line is equal to the resistance.

Later, you learn about V-I curves for other components. They have several uses, and often they are not straight lines.

THE VOLTAGE DIVIDER

23 The circuit shown in Figure 1.7 is called a *voltage divider*. It is the basis for many important theoretical and practical ideas you encounter throughout the entire field of electronics.

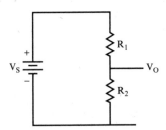

FIGURE 1.7

The object of this circuit is to create an output voltage (V_0) that you can control based upon the two resistors and the input voltage. V_0 is also the *voltage drop* across R_2.

QUESTION

What is the formula for V_0? _____

_____

ANSWER

$$V_o = V_S \times \frac{R_2}{R_1 + R_2}$$

$R_1 + R_2 = R_T$, the total resistance of the circuit.

24 A simple example can demonstrate the use of this formula.

QUESTION

For the circuit shown in Figure 1.8, what is V_0? _____

_____

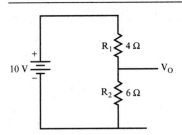

FIGURE 1.8

ANSWER

$$V_O = V_S \times \frac{R_2}{R_1 + R_2}$$

$$= 10 \times \frac{6}{4+6}$$

$$= 10 \times \frac{6}{10}$$

$$= 6 \text{ volts}$$

25 Now, try these problems.

QUESTIONS

What is the output voltage for each combination of supply voltage and resistance?

A. V_S = 1 volt, R_1 = 1 ohm, R_2 = 1 ohm

V_0 = _____

B. V_S = 6 volts, R_1 = 4 ohms, R_2 = 2 ohms

V_0 = _____

C. V_S = 10 volts, R_1 = 3.3. kΩ, R_2 = 5.6 kΩ

V_0 = _____

D. V_S = 28 volts, R_1 = 22 kΩ, R_2 = 6.2 kΩ

V_0 = _____

ANSWERS

A. 0.5 volts

B. 2 volts

C. 6.3 volts

D. 6.16 volts

26 The output voltage from the voltage divider is always less than the applied voltage. Voltage dividers are often used to apply specific voltages to different components in a circuit. Use the voltage divider equation to answer the following questions.

QUESTIONS

A. What is the voltage drop across the 22 kΩ resistor for question D of problem 25? __

_____

B. What total voltage do you get if you add this voltage drop to the voltage drop across the 6.2 kΩ resistor? _____

_____

ANSWERS

A. 21.84 volts

B. The sum is 28 volts.

The voltages across the two resistors add up to the supply voltage. This is an example of *Kirchhoff's Voltage Law (KVL)*, which simply means that the voltage supplied to a circuit must equal the sum of the voltage drops in the circuit. In this book, KVL is often used without actual reference to the law.

Also the voltage drop across a resistor is proportional to the resistor's value. Therefore, if one resistor has a greater value than another in a series circuit, the voltage drop across the higher-value resistor is greater.

USING BREADBOARDS

A convenient way to create a prototype of an electronic circuit to verify that it works is to build it on a *breadboard*. You can use breadboards to build the circuits used in the projects later in this book. As shown in the following figure, a breadboard is a sheet of plastic with several contact holes. You use these holes to connect electronic components in a circuit. After you verify that a circuit works with this method, you can then create a permanent circuit using soldered connections.

Continued

(continued)

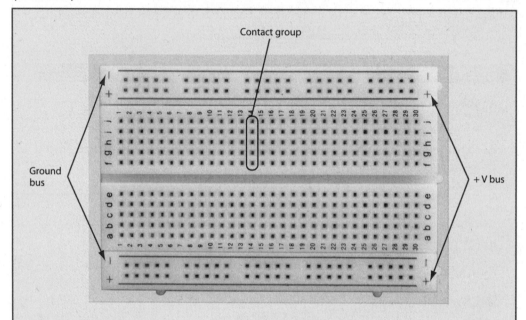

Contact group

Ground bus

+V bus

Breadboards contain metal strips arranged in a pattern under the contact holes, which are used to connect groups of contacts together. Each group of five contact holes in a vertical line (such as the group circled in the figure) is connected by these metal strips. Any components plugged into one of these five contact holes are, therefore, electrically connected.

Each row of contact holes marked by a "+" or "–" are connected by these metal strips. The rows marked "+" are connected to the positive terminal of the battery or power supply and are referred to as the +V bus. The rows marked "–" are connected to the negative terminal of the battery or power supply and are referred to as the *ground bus*. The +V buses and ground buses running along the top and bottom of the breadboard make it easy to connect any component in a circuit with a short piece of wire called a jumper wire. Jumper wires are typically made of 22-gauge solid wire with approximately 1/4 inch of insulation stripped off each end.

The following figure shows a voltage divider circuit assembled on a breadboard. One end of R_1 is inserted into a group of contact holes that is also connected by a jumper wire to the +V bus. The other end of R_1 is inserted into the same group of contact holes that contains one end of R_2. The other end of R_2 is inserted into a group of contact holes that is also connected by a jumper wire to the ground bus. In this example, a 1.5 kΩ resistor was used for R_1, and a 5.1 kΩ resistor was used for R_2.

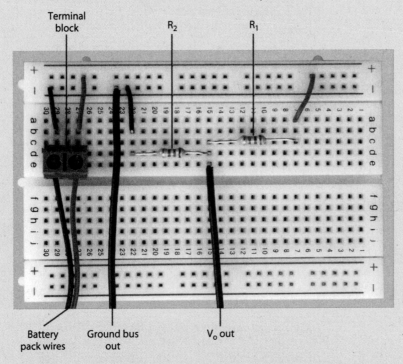

A *terminal block* is used to connect the battery pack to the breadboard because the wires supplied with battery packs (which are stranded wire) can't be inserted directly into breadboard contact holes. The red wire from a battery pack is attached to the side of the terminal block that is inserted into a group of contact holes, which is also connected by a jumper wire to the +V bus. The black wire from a battery pack is attached to the side of the terminal block that is inserted into a group of contact holes, which is also connected by a jumper wire to the ground bus.

To connect the output voltage, V_o, to a multimeter or a downstream circuit, two additional connections are needed. One end of a jumper wire is inserted in the same group of contact holes that contain both R_1 and R_2 to supply V_o. One end of another jumper wire is inserted in a contact hole in the ground bus to provide an electrical contact to the negative side of the battery. When connecting test equipment to the breadboard, you should use a 20-gauge jumper wire because sometimes the 22-gauge wire is pulled out of the board when attaching test probes.

THE CURRENT DIVIDER

27 In the circuit shown in Figure 1.9, the current splits or divides between the two resistors that are connected in parallel.

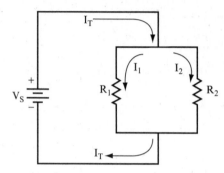

FIGURE 1.9

I_T splits into the individual currents I_1 and I_2, and then these recombine to form I_T.

QUESTIONS

Which of the following relationships are valid for this circuit?

A. $V_S = R_1 I_1$

B. $V_S = R_2 I_2$

C. $R_1 I_1 = R_2 I_2$

D. $I_1/I_2 = R_2/R_1$

ANSWERS

All of them are valid.

28 When solving current divider problems, follow these steps:

1. Set up the ratio of the resistors and currents:

$I_1/I_2 = R_2/R_1$

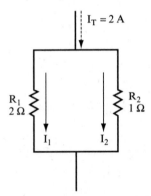

2. Rearrange the ratio to give I_2 in terms of I_1:

$$I_2 = I_1 \times \frac{R_1}{R_2}$$

3. From the fact that $I_T = I_1 + I_2$, express I_T in terms of I_1 only.

4. Now, find I_1.

5. Now, find the remaining current (I_2).

QUESTION

The values of two resistors in parallel and the total current flowing through the circuit are shown in Figure 1.10. What is the current through each individual resistor?

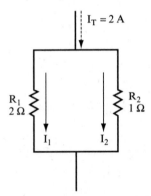

FIGURE 1.10

ANSWER

Work through the steps as shown here:

1. $I_1/I_2 = R_2/R_1 = 1/2$

2. $I_2 = 2I_1$

3. $I_T = I_1 + I_2 = I_1 + 2I_1 = 3I_1$

4. $I_1 = I_T/3 = 2/3$ ampere

5. $I_2 = 2I_1 = 4/3$ amperes

29 Now, try these problems. In each case, the total current and the two resistors are given. Find I_1 and I_2.

QUESTIONS

A. $I_T = 30$ mA, $R_1 = 12$ kΩ, $R_2 = 6$ kΩ _____

_____

B. $I_T = 133$ mA, $R_1 = 1$ kΩ, $R_2 = 3$ kΩ _____

_____

C. What current do you get if you add I_1 and I_2? _____

_____

ANSWERS

A. $I_1 = 10$ mA, $I_2 = 20$ mA

B. $I_1 = 100$ mA, $I_2 = 33$ mA

C. They add back together to give you the total current supplied to the parallel circuit.

Question C is actually a demonstration of *Kirchhoff's Current Law (KCL)*. Simply stated, this law says that the total current entering a junction in a circuit must equal the sum of the currents leaving that junction. This law is also used on numerous occasions in later chapters. KVL and KCL together form the basis for many techniques and methods of analysis that are used in the application of circuit analysis.

Also, the current through a resistor is inversely proportional to the resistor's value. Therefore, if one resistor is larger than another in a parallel circuit, the current flowing through the higher value resistor is the smaller of the two. Check your results for this problem to verify this.

30 You can also use the following equation to calculate the current flowing through a resistor in a two-branch parallel circuit:

$$I_1 = \frac{(I_T)(R_2)}{(R_1 + R_2)}$$

QUESTION

Write the equation for the current I_2. _____

_____

Check the answers for the previous problem using these equations.

ANSWER

$$I_2 = \frac{(I_T)(R_2)}{(R_1 + R_2)}$$

The current through one branch of a two-branch circuit is equal to the total current times the resistance of the opposite branch, divided by the sum of the resistances of both branches. This is an easy formula to remember.

USING THE MULTIMETER

A *multimeter* is a must-have testing device for anyone's electronics toolkit. A multimeter is aptly named because it can be used to measure multiple parameters. Using a multimeter, you can measure current, voltage, and resistance by setting the rotary switch on the multimeter to the parameter you want to measure, and connecting each mulitmeter probe to a wire in a circuit. The following figure shows a multimeter connected to a voltage divider circuit to measure voltage. Following are the details of how you take each of these measurements.

VOLTAGE

To measure the voltage in the circuit shown in the figure, at the connection between R_1 and R_2, use jumper wire to connect the red probe of a multimeter to the row of contact holes containing leads from both R_1 and R_2. Use another jumper wire to connect the black probe of the multimeter to the ground bus. Set the rotary switch on the multimeter to measure voltage, and it returns the results.

Continued

(continued)

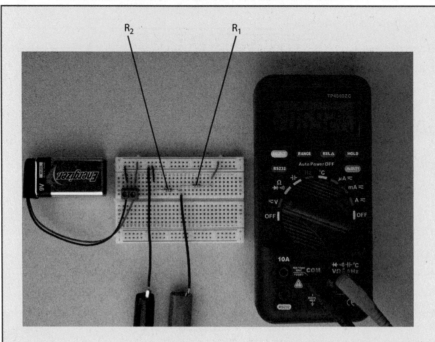

NOTE The circuit used in a multimeter to measure voltage places a large-value resistor in parallel with R_2 so that the test itself does not cause any measurable drop in the current passing through the circuit.

TIP Whenever you perform tests on a circuit, attach alligator clips or test clips with plastic covers to the ends of the probes. This aids the probes in grabbing the jumper wires with little chance that they'll cause a short.

CURRENT

The following figure shows how you connect a multimeter to a voltage divider circuit to measure current. Connect a multimeter in series with components in the circuit, and set the rotary switch to the appropriate ampere range, depending upon the magnitude of the expected current. To connect the multimeter in series with R_1 and R_2, use a jumper wire to connect the +V bus to the red lead of a multimeter, and another jumper wire to connect the black lead of the multimeter to R_1. These connections force the current flowing through the circuit to flow through the multimeter.

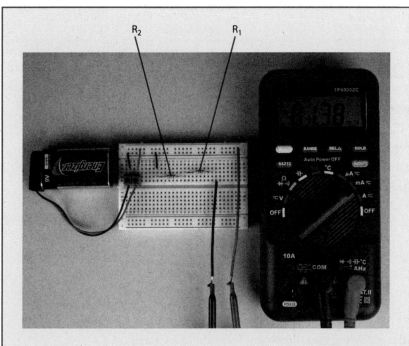

NOTE   The circuit used in a multimeter to measure current passes the current through a low-value resistor so that the test itself does not cause any measurable drop in the current.

RESISTANCE

You typically use the resistance setting on a multimeter to check the resistance of individual components. For example, in measuring the resistance of R_2 before assembling the circuit shown in the previous figure, the result was 5.0 kΩ, slightly off the nominal 5.1 kΩ stated value.

You can also use a multimeter to measure the resistance of a component in a circuit. A multimeter measures resistance by applying a small current through the components being tested, and measuring the voltage drop. Therefore, to prevent false readings, you should disconnect the battery pack or power supply from the circuit before using the multimeter.

SWITCHES

31 A *mechanical switch* is a device that completes or breaks a circuit. The most familiar use is that of applying power to turn a device on or off. A switch can also permit a signal to pass from one place to another, prevent its passage, or route a signal to one of several places.

In this book, you work with two types of switches. The first is the simple on-off switch, also called a *single pole single throw* switch. The second is the *single pole double throw* switch. Figure 1.11 shows the circuit symbols for each.

ON-OFF switch

in the OFF position

Single pole double throw or SPDT switch

FIGURE 1.11

Keep in mind the following two important facts about switches:

- A closed (or ON) switch has the total circuit current flowing through it. There is no voltage drop across its terminals.

- An open (or OFF) switch has no current flowing through it. The full circuit voltage appears between its terminals.

The circuit shown in Figure 1.12 includes a closed switch.

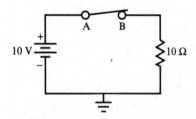

10 V   A   B   10 Ω

FIGURE 1.12

QUESTIONS

A. What is the current flowing through the switch? _____

_____

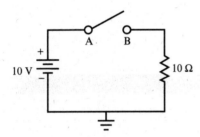

B. What is the voltage at point A and point B with respect to ground? _____

_____

C. What is the voltage drop across the switch? _____

_____

ANSWERS

A. $\dfrac{10\ \text{volts}}{10\ \text{ohms}} = 1\ \text{ampere}$

B. $V_A = V_B = 10\ \text{volts}$

C. 0 V (There is no voltage drop because both terminals are at the same voltage.)

32 The circuit shown in Figure 1.13 includes an open switch.

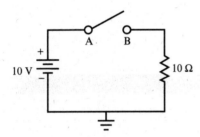

FIGURE 1.13

QUESTIONS

A. What is the voltage at point A and point B? _____

_____

B. How much current is flowing through the switch? _____

_____

C. What is the voltage drop across the switch? _____

_____

ANSWERS

A. $V_A = 10$ volts; $V_B = 0$ volts.

B. No current is flowing because the switch is open.

C. 10 volts. If the switch is open, point A is the same voltage as the positive battery terminal, and point B is the same voltage as the negative battery terminal.

33 The circuit shown in Figure 1.14 includes a single pole double throw switch. The position of the switch determines whether lamp A or lamp B is lit.

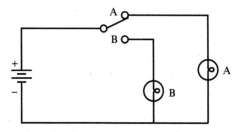

FIGURE 1.14

QUESTIONS

A. In the position shown, which lamp is lit? _____

_____

B. Can both lamps be lit simultaneously? _____

_____

ANSWERS

A. Lamp A.

B. No, one or the other must be off.

CAPACITORS IN A DC CIRCUIT

34 Capacitors are used extensively in electronics. They are used in both alternating current (AC) and DC circuits. Their main use in DC electronics is to become charged, hold the charge, and, at a specific time, release the charge.

The capacitor shown in Figure 1.15 charges when the switch is closed.

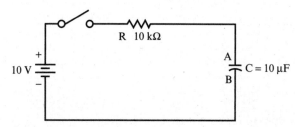

FIGURE 1.15

QUESTION

To what final voltage will the capacitor charge? _____

_____

ANSWER

It will charge up to 10 volts. It will charge up to the voltage that would appear across an open circuit located at the same place where the capacitor is located.

35 How long does it take to reach this voltage? This is an important question with many practical applications. To find the answer you must know the *time constant (τ)* (Greek letter tau) of the circuit.

QUESTIONS

A. What is the formula for the time constant of this type of circuit? _____

_____

B. What is the time constant for the circuit shown in Figure 1.15? _____

_____

C. How long does it take the capacitor to reach 10 volts? _____

_____

D. To what voltage level does it charge in one time constant? _____

_____

ANSWERS

A. $\tau = R \times C$.

B. $\tau = 10\,\text{k}\Omega \times 10\,\mu\text{F} = 10{,}000\,\Omega \times 0.00001\,\text{F} = 0.1$ seconds. (Convert resistance values to ohms and capacitance values to farads for this calculation.)

C. Approximately 5 time constants, or about 0.5 seconds.

D. 63 percent of the final voltage, or about 6.3 volts.

---

36 The capacitor does not begin charging until the switch is closed. When a capacitor is uncharged or discharged, it has the same voltage on both plates.

QUESTIONS

A. What is the voltage on plate A and plate B of the capacitor in Figure 1.15 before the switch is closed? _____

_____

B. When the switch is closed, what happens to the voltage on plate A? _____

_____

C. What happens to the voltage on plate B? _____

_____

D. What is the voltage on plate A after one time constant? _____

_____

ANSWERS

A. Both will be at 0 volts if the capacitor is totally discharged.

B. It will rise toward 10 volts.

C. It will stay at 0 volts.

D. About 6.3 volts.

37 The *capacitor charging graph* in Figure 1.16 shows how many time constants a voltage must be applied to a capacitor before it reaches a given percentage of the applied voltage.

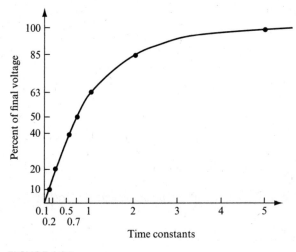

FIGURE 1.16

QUESTIONS

A. What is this type of curve called? _____

_____

B. What is it used for? _____

_____

ANSWERS

A. It is called an *exponential* curve.

B. It is used to calculate how far a capacitor has charged in a given time.

38 In the following examples, a resistor and a capacitor are in series. Calculate the time constant, how long it takes the capacitor to fully charge, and the voltage level after one time constant if a 10-volt battery is used.

QUESTIONS

A. $R = 1 \text{ k}\Omega$, $C = 1,000 \text{ }\mu\text{F}$ _____

_____

B. $R = 330 \text{ k}\Omega$, $C = 0.05 \text{ }\mu\text{F}$ _____

_____

ANSWERS

A. $\tau = 1$ second; charge time = 5 seconds; $V_C = 6.3$ volts.

B. $\tau = 16.5$ ms; charge time = 82.5 ms; $V_C = 6.3$ volts. (The abbreviation "ms" indicates milliseconds.)

39 The circuit shown in Figure 1.17 uses a double pole switch to create a discharge path for the capacitor.

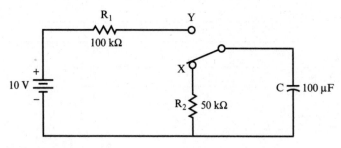

FIGURE 1.17

QUESTIONS

A. With the switch in position X, what is the voltage on each plate of the capacitor? __

_____

B. When the switch is moved to position Y, the capacitor begins to charge. What is its charging time constant? _____

_____

C. How long does it take to fully charge the capacitor? _____

_____

ANSWERS

A. 0 volts

B. $\tau = R \times C = (100 \text{ k}\Omega)(100 \text{ }\mu\text{F}) = 10 \text{ secs}$

C. Approximately 50 seconds

40 Suppose that the switch shown in Figure 1.17 is moved back to position X after the capacitor is fully charged.

QUESTIONS

A. What is the discharge time constant of the capacitor? _____

_____

B. How long does it take to fully discharge the capacitor? _____

_____

ANSWERS

A. $\tau = R \times C = (50 \text{ k}\Omega)(100 \text{ }\mu\text{F}) = 5 \text{ seconds}$ (discharging through the 50 kΩ resistor)

B. Approximately 25 seconds

Continued

(continued)

ANSWERS

The circuit powering a camera flash is an example of a capacitor's capability to store a charge and then discharge upon demand. While you wait for the flash unit to charge, the camera uses its battery to charge a capacitor. When the capacitor is charged, it holds that charge until you click the Shutter button, causing the capacitor to discharge, which powers the flash.

INSIDE THE CAPACITOR

Capacitors store an electrical charge on conductive plates that are separated by an insulating material, as shown in the following figure. One of the most common types of capacitor is a *ceramic capacitor*, which has values ranging from a few μF up to approximately 47 μF. The name for a ceramic capacitor comes from the use of a ceramic material to provide insulation between the metal plates.

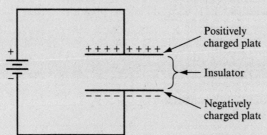

Another common type of capacitor is an electrolytic capacitor, available with capacitance values ranging from 0.1 μF to several thousand μF. The name electrolytic comes from the use of an electrolytic fluid, which, because it is conductive, acts as one of the "plates," whereas the other plate is made of metal. The insulating material is an oxide on the surface of the metal.

Unlike ceramic capacitors, many electrolytic capacitors are polarized, which means that you must insert the lead marked with a "+" in the circuit closest to the positive voltage source. The symbol for a capacitor indicates the direction in which you insert polarized capacitors in a circuit. The curved side of the capacitor symbol indicates the negative side of the capacitor, whereas the straight side of the symbol indicates the positive side of the capacitor. You can see this orientation later in this chapter in Figure 1.22.

Units of capacitance are stated in pF (picofarad), μF (microfarad), and F (farad). One μF equals 1,000,000 pF and one F equals 1,000,000 μF. Many capacitors are marked with their capacitance value, such as 220 pF. However, you'll often find capacitors that use a different numerical code, such as 224. The first two numbers in this code are the first and second significant digits of the capacitance value. The third number is the multiplier, and the units are pF. Therefore, a capacitor marked with 221 has a value of 220 pF, whereas a capacitor with a marking of 224 has a value of 220,000 pF. (You can simplify this to 0.22 μF.)

41 Capacitors can be connected in parallel, as shown in Figure 1.18.

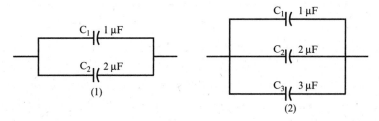

FIGURE 1.18

QUESTIONS

A. What is the formula for the total capacitance? _____

_____

B. What is the total capacitance in circuit 1? _____

_____

C. What is the total capacitance in circuit 2? _____

_____

ANSWERS

A. $C_T = C_1 + C_2 + C_3 + \ldots + C_N$

B. $C_T = 1 + 2 = 3\ \mu F$

C. $C_T = 1 + 2 + 3 = 6\ \mu F$

In other words, the total capacitance is found by simple addition of the capacitor values.

42 Capacitors can be placed in series, as shown in Figure 1.19.

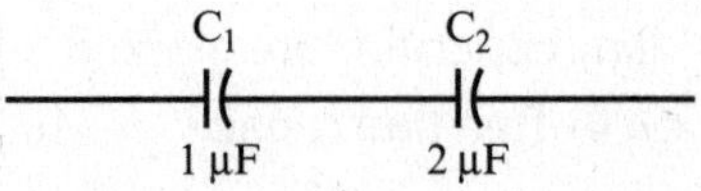

FIGURE 1.19

QUESTIONS

A. What is the formula for the total capacitance? _____

_____

B. In Figure 1.19, what is the total capacitance? _____

_____

ANSWERS

A. $\dfrac{1}{C_T} = \dfrac{1}{C_1} + \dfrac{1}{C_2} + \dfrac{1}{C_3} + \cdots + \dfrac{1}{C_N}$

B. $\dfrac{1}{C_T} = \dfrac{1}{1} + \dfrac{1}{2} = 1 + \dfrac{1}{2} = \dfrac{3}{2}$; thus $C_T = \dfrac{2}{3}$

43 In each of these examples, the capacitors are placed in series. Find the total capacitance.

QUESTIONS

A. $C_1 = 10\ \mu F$, $C_2 = 5\ \mu F$ _____

_____

B. $C_1 = 220\ \mu F$, $C_2 = 330\ \mu F$, $C_3 = 470\ \mu F$ _____

_____

C. $C_1 = 0.33\ \mu F$, $C_2 = 0.47\ \mu F$, $C_3 = 0.68\ \mu F$ _____

_____

ANSWERS

A. 3.3 µF

B. 103.06 µF

C. 0.15 µF

SUMMARY

The few simple principles reviewed in this chapter are those you need to begin the study of electronics. Following is a summary of these principles:

- The basic electrical circuit consists of a source (voltage), a load (resistance), and a path (conductor or wire).

- The voltage represents a charge difference.

- If the circuit is a complete circuit, then electrons flow, which is called current flow. The resistance offers opposition to current flow.

- The relationship between V, I, and R is given by Ohm's law:

$$V = I \times R$$

- Resistance could be a combination of resistors in series, in which case you add the values of the individual resistors together to get the total resistance.

$$R_T = R_1 + R_2 + \cdots + R_N$$

- Resistance can be a combination of resistors in parallel, in which case you find the total by using the following formula:

$$\frac{1}{R_T} = \frac{1}{R_1} + \frac{1}{R_2} + \frac{1}{R_3} + \cdots + \frac{1}{R_N} \quad \text{or} \quad R_T = \frac{1}{\dfrac{1}{R_1} + \dfrac{1}{R_2} + \dfrac{1}{R_3} + \cdots + \dfrac{1}{R_N}}$$

- You can find the power delivered by a source by using the following formula:

$$P = VI$$

- You can find the power dissipated by a resistance by using the following formula:

$$P = I^2R \quad \text{or} \quad P = \frac{V^2}{R}$$

- If you know the total applied voltage, V_s, you can find the voltage across one resistor in a series string of resistors by using the following voltage divider formula:

$$V_1 = \frac{V_S R_1}{R_T}$$

- You can find the current through one resistor in a two resistor parallel circuit with the total current known by using the current divider formula:

$$I_1 = \frac{I_T R_2}{(R_1 + R_2)}$$

- Kirchhoff's Voltage Law (KVL) relates the voltage drops in a series circuit to the total applied voltage.

$$V_S = V_1 + V_2 + \cdots + V_N$$

- Kirchhoff's Current Law (KCL) relates the currents at a junction in a circuit by saying that the sum of the input currents equals the sum of the output currents. For a simple parallel circuit, this becomes the following, where I_T is the input current:

$$I_T = I_1 + I_2 + \cdots + I_N$$

- A switch in a circuit is the control device that directs the flow of current or, in many cases, allows that current to flow.

- Capacitors are used to store electric charge in a circuit. They also allow current or voltage to change at a controlled pace. The circuit time constant is found by using the following formula:

$$\tau = R \times C$$

- At one time constant in an RC circuit, the values for current and voltage have reached 63 percent of their final values. At five time constants, they have reached their final values.

- Capacitors in parallel are added to find the total capacitance.

$$C_T = C_1 + C_2 + \cdots + C_N$$

- Capacitors in series are treated the same as resistors in parallel to find a total capacitance.

$$\frac{1}{C_T} = \frac{1}{C_1} + \frac{1}{C_2} + \cdots + \frac{1}{C_N} \quad \text{or} \quad C_T = \frac{1}{\dfrac{1}{C_1} + \dfrac{1}{C_2} + \dfrac{1}{C_3} + \cdots + \dfrac{1}{C_N}}$$

DC PRE-TEST

The following problems and questions test your understanding of the basic principles presented in this chapter. You need a separate sheet of paper for your calculations. Compare your answers with the answers provided following the test. You can work many of the problems in more than one way.

Questions 1–5 use the circuit shown in Figure 1.20. Find the unknown values indicated using the values given.

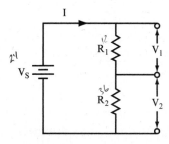

FIGURE 1.20

1. R_1 = 12 ohms, R_2 = 36 ohms, V_S = 24 volts

   R_T = ___48___ , I = ___0.5___

2. R_1 = 1 kΩ, R_2 = 3 kΩ, I = 5 mA

   V_1 = ___5___ , V_2 = ___15___ , V_S = ___26___

3. $R_1 = 12$ kΩ, $R_2 = 8$ kΩ, $V_S = 24$ volts

$V_1 =$ _____14.4_____ , $V_2 =$ _9.6_

4. $V_S = 36$ V, I $= 250$ mA, $V_1 = 6$ volts

$R_2 =$ _36-6=30v ÷ .250 = 120 Ω_

5. Now, go back to problem 1. Find the power dissipated by each resistor and the total power delivered by the source.

$P_1 =$ _6x.5=3w_ , $P_2 =$ _18x.5=9w_ , $P_T =$ _12w_

Questions 6–8 use the circuit shown in Figure 1.21. Again, find the unknowns using the given values.

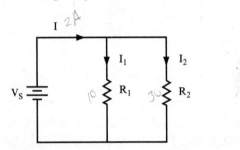

FIGURE 1.21

6. $R_1 = 6$ kΩ, $R_2 = 12$ kΩ, $V_S = 20$ volts

$R_T =$ _4109_ , I $=$ _0.25 mA_

7. I $= 2$ A, $R_1 = 10$ ohms, $R_2 = 30$ ohms

$I_1 =$ _____ , $I_2 =$ _____

8. $V_S = 12$ volts, I $= 300$ mA, $R_1 = 50$ ohms

$R_2 =$ _____ , $P_1 =$ _____

9. What is the maximum current that a 220- ohm resistor can safely have if its power rating is 1/4 watt?

$I_{MAX} =$ _____

10. In a series RC circuit the resistance is 1 kΩ, the applied voltage is 3 volts, and the time constant should be 60 μsec.

A. What is the required value of C?

C $=$ _____

B. What is the voltage across the capacitor 60 μsec after the switch is closed?

$V_C = $ _____

C. At what time will the capacitor be fully charged?

T = _____

11. In the circuit shown in Figure 1.22, when the switch is at position 1, the time constant should be 4.8 ms.

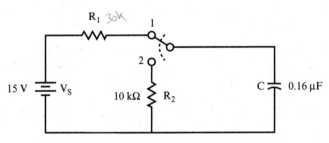

FIGURE 1.22

A. What should be the value of resistor R_1?

$R_1 = $ _30K_____

B. What will be the voltage on the capacitor when it is fully charged, and how long will it take to reach this voltage?

$V_C = $ _15_____ , T = _24ms_____

C. After the capacitor is fully charged, the switch is thrown to position 2. What is the discharge time constant, and how long will it take to completely discharge the capacitor?

τ = ___1.6ms_____ , T = ___8ms_____

12. Three capacitors are available with the following values:

$C_1 = 8$ μF; $C_2 = 4$ μF; $C_3 = 12$ μF.

A. What is C_T if all three are connected in parallel?

$C_T = $ _24uf_____

B. What is C_T if they are connected in series?

$C_T = $ _____2.18uf_____

C. What is C_1 if C_1 is in series with the parallel combination of C_2 and C_3?

$C_T = $ _____5.3uf_____

ANSWERS TO DC PRE-TEST

If your answers do not agree with those provided here, review the problems indicated in parentheses before you go to Chapter 2, "The Diode." If you still feel uncertain about these concepts, go to a website such as www.BuildingGadgets.com and work through DC tutorials listed there.

It is assumed that Ohm's law is well known, so problem 4 will not be referenced.

1.	R_T = 48 ohms, I = 0.5 ampere	(problem 9)
2.	V_1 = 5 volts, V_2 = 15 volts, V_s = 20 volts	(problems 23 and 26)
3.	V_1 = 14.4 volts, V_2 = 9.6 volts	(problems 23 and 26)
4.	R_2 = 120 ohms	(problems 9 and 23)
5.	P_1 = 3 watts, P_2 = 9 watts, P_T = 12 watts	(problems 9 and 13)
6.	R_T = 4 kΩ, I = 5 mA	(problem 10)
7.	I_1 = 1.5 amperes, I_2 = 0.5 ampere	(problems 28 and 29)
8.	R_2 = 200 ohms, P_1 = 2.88 watts	(problems 10 and 13)
9.	I_{MAX} = 33.7 mA	(problems 13, 15, and 16)
10A.	C = 0.06 μF	(problems 34 and 35)
10B.	V_C = 1.9 volts	(problem 35)
10C.	T = 300 μsec	(problems 34–38)
11A.	R_1 = 30 kΩ	(problems 33, 39, and 40)
11B.	V_C = 15 V, T = 24 ms	(problem 35)
11C.	τ = 1.6 ms, T = 8.0 ms	(problems 39–40)
12A.	C_T = 24 μF	(problems 41 and 42)
12B.	C_T = 2.18 μF	(problem 42)
12C.	C_T = 5.33 μF	(problems 42–43)

2

The Diode

The main characteristic of a diode is that it conducts electricity in one direction only. Historically, the first vacuum tube was a diode; it was also known as a *rectifier*. The modern *diode* is a semiconductor device. It is used in all applications where the older vacuum tube diode was used, but it has the advantages of being much smaller, easier to use, and less expensive.

A *semiconductor* is a crystalline material that, depending on the conditions, can act as a conductor (allowing the flow of electric current) or an insulator (preventing the flow of electric current). Techniques have been developed to customize the electrical properties of adjacent regions of semiconductor crystals, which allow the manufacture of small diodes, as well as transistors and integrated circuits.

When you complete this chapter, you can do the following:

- Specify the uses of diodes in DC circuits.
- Determine from a circuit diagram whether a diode is forward- or reverse-biased.
- Recognize the characteristic V-I curve for a diode.
- Specify the knee voltage for a silicon or a germanium diode.
- Calculate current and power dissipation in a diode.
- Define diode breakdown.
- Differentiate between zeners and other diodes.
- Determine when a diode can be considered "perfect."

UNDERSTANDING DIODES

1 Silicon and germanium are semiconductor materials used in the manufacture of diodes, transistors, and integrated circuits. Semiconductor material is refined to an extreme level of purity, and then minute, controlled amounts of a specific impurity are added (a process called *doping*). Based on which impurity is added to a region of a semiconductor crystal, that region is said to be *N type* or *P type*. In addition to electrons (which are negative charge carriers used to conduct charge in a conventional conductor), semiconductors contain positive charge carriers called *holes*. The impurities added to an N type region increases the number of electrons capable of conducting charge, whereas the impurities added to a P type region increase the number of holes capable of conducting charge.

When a semiconductor chip contains an N doped region adjacent to a P doped region, a *diode junction* (often called a *PN junction*) is formed. Diode junctions can also be made with either silicon or germanium. However, silicon and germanium are never mixed when making PN junctions.

QUESTION

Which diagrams in Figure 2.1 show diode junctions? _____

_____

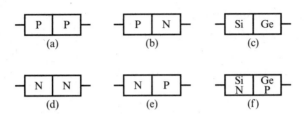

FIGURE 2.1

2 In a diode, the P material is called the *anode*. The N material is called the *cathode*.

QUESTION

Identify which part of the diode shown in Figure 2.2 is P material and which part is N material. _____

_____

Anode Cathode

FIGURE 2.2

3 Diodes are useful because electric current can flow through a PN junction *in one direction only*. Figure 2.3 shows the direction in which the current flows.

FIGURE 2.3

Figure 2.4 shows the circuit symbol for a diode. The arrowhead points in the direction of current flow. Although the anode and cathode are indicated here, they are not usually indicated in circuit diagrams.

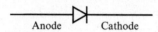

Anode   Cathode

FIGURE 2.4

QUESTION

In a diode, does current flow from anode to cathode, or cathode to anode? _____

_____

ANSWER

Current flows from anode to cathode.

4 In the circuit shown in Figure 2.5, an arrow shows the direction of current flow.

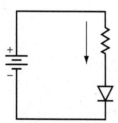

FIGURE 2.5

QUESTIONS

A. Is the diode connected correctly to permit current to flow? _____

_____

B. Notice the way the battery and the diode connect. Is the anode at a higher or lower voltage than the cathode? _____

_____

ANSWERS

A. Yes.

B. The anode connects to the positive battery terminal, and the cathode connects to the negative battery terminal. Therefore, the anode is at a higher voltage than the cathode.

5 When the diode is connected so that the current flows, it is *forward-biased*. In a forward-biased diode, the anode connects to a higher voltage than the cathode, and current flows. Examine the way the diode is connected to the battery shown in Figure 2.6.

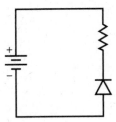

FIGURE 2.6

QUESTION

Is the diode forward-biased? Give the reasons for your answer. _____

_____

ANSWER

No, it is not forward-biased. The cathode is connected to the positive battery terminal, and the anode is connected to the negative battery terminal. Therefore, the cathode is at a higher voltage than the anode.

6 When the cathode is connected to a higher voltage level than the anode, the diode cannot conduct. In this case, the diode is *reverse-biased*.

QUESTION

Draw a reverse-biased diode in the circuit shown in Figure 2.7. _____

_____

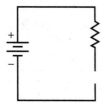

FIGURE 2.7

ANSWER

Your drawing should look something like Figure 2.8.

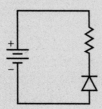

FIGURE 2.8

7 In many circuits, the diode is often considered to be a *perfect* diode to simplify calculations. A perfect diode has zero voltage drop in the forward direction and conducts no current in the reverse direction.

QUESTION

From your knowledge of basic electricity, what other component has zero voltage drop across its terminals in one condition and conducts no current in an alternative condition?

_____

ANSWER

The mechanical switch. When closed, it has no voltage drop across its terminals, and when open, it conducts no current.

8 A forward-biased perfect diode can thus be compared to a closed switch. It has no voltage drop across its terminals, and current flows through it.

A reverse-biased perfect diode can be compared to an open switch. No current flows through it, and the voltage difference between its terminals equals the supply voltage.

QUESTION

Which of the switches shown in Figure 2.9 performs like a forward-biased perfect diode? _____

_____

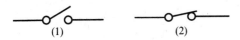

        (1)   (2)

FIGURE 2.9

ANSWER

Switch (2) represents a closed switch and, like a forward-biased perfect diode, allows current to flow through it. There is no voltage drop across its terminals.

PROJECT 2.1: The Diode

If you have access to electronic equipment, you may want to perform the simple project described in the next few problems. If this is the first time you have tried such a project, get help from an instructor or someone who is familiar with electronic projects.

When building electronic circuits, eventually you'll make a mistake (as all of us do), and sometimes those mistakes cause circuits to fry. If you smell hot electronic components, disconnect the battery from the circuit, and then check the circuit to determine what connections you should change.

When fixing a circuit, follow some simple safety rules. Do not try to rearrange connections with the battery connected because you may short leads together.

Also, don't touch bare wires with live electricity. Even with batteries, you have a chance of being burned or seriously injured. If your skin is wet, it forms an electrical connection with lower resistance, allowing more current to flow, potentially injuring you.

If you do not have access to equipment, do not skip this project. Read through the project, and try to picture or imagine the results. This is sometimes called "dry-labbing" the experiment. You can learn a lot from reading about this project, even though it is always better to actually perform the project. This advice also applies to the other projects in many of the following chapters.

OBJECTIVE

The objective of this project is to plot the V-I curve (also called a *characteristic curve*) of a diode, which shows how current flow through the diode varies with the applied voltage. As shown in Figure 2.10, the I-V curve for a diode demonstrates that if low voltage is applied to a diode, current does not flow. However, when the applied voltage exceeds a certain value, the current flow increases quickly.

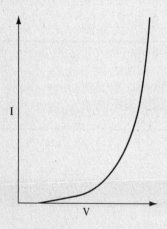

FIGURE 2.10

GENERAL INSTRUCTIONS

While the circuit is set up, measure the current for each voltage value. As you perform the project, observe how much more rapidly the current increases for higher applied voltages.

Parts List

- ❑ One 9 V battery
- ❑ One snap battery connector
- ❑ One multimeter set to measure current (mA)
- ❑ One multimeter set to measure DC voltage

❑ One 330-ohm, 0.5-watt resistor

❑ One 1N4001 diode

❑ One breadboard

❑ One 1 MΩ potentiometer

❑ One terminal block

STEP-BY-STEP INSTRUCTIONS

Set up the circuit as shown in Figure 2.11. The circled "A" designates a multimeter set to measure current, and the circled "V" designates a multimeter set to measure DC voltage. If you have some experience in building circuits, this schematic (along with the previous parts list) should provide all the information you need to build the circuit. If you need a bit more help in building the circuit, look at the photos of the completed circuit in the "Expected Results" section.

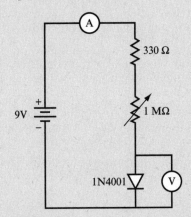

FIGURE 2.11

Carefully check your circuit against Figure 2.11, especially the direction of the battery and the diode. The diode has a band at one end. Connect the lead at the end of the diode with the band to the ground bus on the breadboard.

After you check your circuit, follow these steps, and record your measurements in the blank table following the steps:

1. Set the potentiometer to its highest value. This sets the voltage applied to the diode to its lowest possible value.

2. Measure and record the voltage applied to the diode.

3. Measure and record the current.

4. Adjust the potentiometer slightly to give a higher voltage.

5. Measure and record the new values of voltage and current.

6. Repeat steps 4 and 5 until the lowest resistance of the potentiometer is reached, taking as many readings as possible. This results in the highest voltage and current readings for this circuit. At this point, the potentiometer resistance is zero ohms, and the current is limited to approximately 27 mA by the 330-ohm resistor. This resistor is included in the circuit to avoid overheating the components when the potentiometer is set to zero ohms. If your circuit allows currents significantly above this level as you adjust the potentiometer, something is wrong. You should disconnect the battery and examine the circuit to see if it were connected incorrectly. If V gets large—above 3 or 4 volts—and I remains small, then the diode is backward. Reverse it and start again.

V (volts)	I (mA)

7. Graph the points recorded in the table using the blank graph shown in Figure 2.12. Your curve should look like the one shown in Figure 2.10.

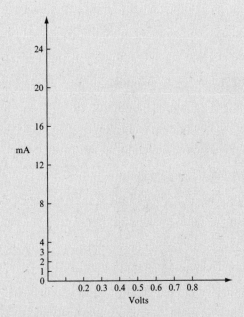

FIGURE 2.12

EXPECTED RESULTS

Figure 2.13 shows the breadboarded circuit for this project.

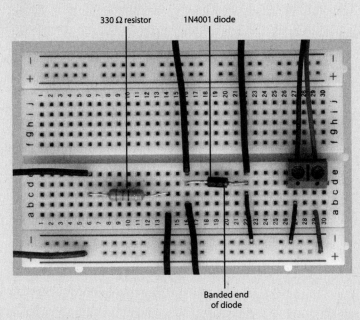

FIGURE 2.13

Figure 2.14 shows the test setup for this project.

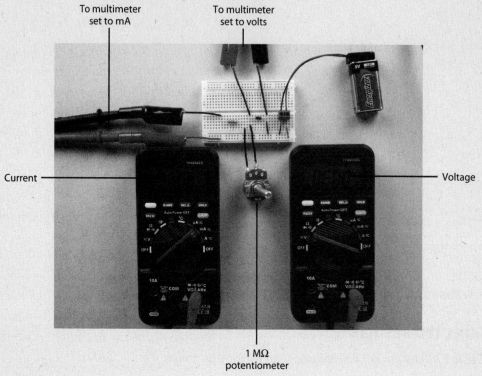

FIGURE 2.14

Compare your measurements with the ones shown in the following table:

V (volts)	I (mA)
0.44	0.00
0.46	0.01
0.50	0.06
0.52	0.11
0.55	0.23
0.58	0.49
0.60	0.92
0.63	1.74
0.68	4.86
0.72	15.1
0.73	20.9
0.74	25.2

Further reductions in the resistance below the 330 Ω included in the circuit causes little increase in the voltage but produces large increases in the current.

Figure 2.15 shows the V-I curve generated using the measurements shown in the preceding table.

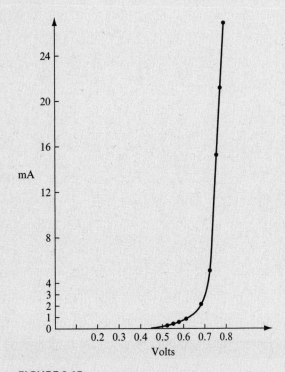

FIGURE 2.15

The V-I curve (or diode characteristic curve) is repeated in Figure 2.16 with three important regions marked on it.

The most important region is the *knee region*. This is not a sharply defined changeover point, but it occupies a narrow range of the curve where the diode resistance changes from high to low.

The ideal curve is shown for comparison.

For the diode used in this project, the knee voltage is about 0.7 volt, which is typical for a silicon diode. This means (and your data should verify this) that at voltage levels below 0.7 volt, the diode has such a high resistance that it limits the current flow to a low value. This characteristic knee voltage is sometimes referred to as a *threshold voltage*. If you use a germanium diode, the knee voltage is approximately 0.3 volt.

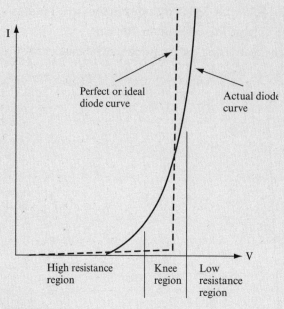

I

Perfect or ideal
diode curve

Actual diode
curve

V

High resistance
region

Knee
region

Low
resistance
region

FIGURE 2.16

9 The knee voltage is also a *limiting voltage*. That is, it is the highest voltage that can be obtained across the diode in the forward direction.

QUESTIONS

A. Which has the higher limiting voltage, germanium or silicon? _____

_____

B. What happens to the diode resistance at the limiting or knee voltage? _____

_____

ANSWERS

A. Silicon, with a limiting voltage of 0.7 volt, is higher than germanium, which has a limiting voltage of only 0.3 volt.

B. It changes from high to low.

NOTE   You use these knee voltages in many later chapters as the voltage drop across the PN junction when it is forward-biased.

10 Refer back to the diagram of resistance regions shown in Figure 2.16.

QUESTION

What happens to the current when the voltage becomes limited at the knee? _____

ANSWER

It increases rapidly.

---

11 For any given diode, the knee voltage is not exactly 0.7 volt or 0.3 volt. Rather, it varies slightly. But when using diodes in practice (that is, imperfect diodes), you can make two assumptions:

√ ▪ The voltage drop across the diode is either 0.7 volt or 0.3 volt.

▪ You can prevent excessive current from flowing through the diode by using the appropriate resistor in series with the diode.

QUESTIONS

A. Why are imperfect diodes specified here? _____

_____

B. Would you use a high or low resistance to prevent excessive current? _____

_____

ANSWERS

A. All diodes are imperfect, and the 0.3 or 0.7 voltage values are only approximate. In fact, in some later problems, it is assumed that the voltage drop across the diode, when it is conducting, is 0 volts. This assumes, then, that as soon as you apply any voltage above 0, current flows in an ideal resistor. (That is, the knee voltage on the V-I curve for an ideal diode is 0 volts.)

√ **B.** Generally, use a high resistance. However, you can calculate the actual resistance value given the applied voltage and the maximum current the diode can withstand.

12 Calculate the current through the diode in the circuit shown in Figure 2.17 using the steps in the following questions.

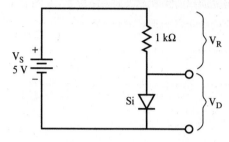

FIGURE 2.17

QUESTIONS

A. The voltage drop across the diode is known. It is 0.7 volt for silicon and 0.3 volt for germanium. ("Si" near the diode means it is silicon.) Write down the diode voltage drop.

$V_D =$ _____

B. Find the voltage drop across the resistor. This can be calculated using $V_R = V_S - V_D$. This is taken from KVL, which was discussed in Chapter 1, "DC Review and Pre-Test."

$V_R =$ _____

C. Calculate the current through the resistor. Use $I = V_R/R$.

$I =$ _____

D. Finally, determine the current through the diode.

$I =$ _____

ANSWERS

You should have written these values:

A. 0.7 volt

B. $V_R = V_S - V_D = 5$ volts $- 0.7$ volt $= 4.3$ volts

C. $I = \dfrac{V_R}{R} = \dfrac{4.3\,\text{volts}}{1\text{k}\Omega} = 4.3\,\text{mA}$

D. 4.3 mA (In a series circuit, the same current runs through each component.)

13 In practice, when the battery voltage is 10 volt or above, the voltage drop across the diode is often considered to be 0 volt, instead of 0.7 volt.

The assumption here is that the diode is a perfect diode, and the knee voltage is at 0 volts, rather than at a threshold value that must be exceeded. As discussed later, this assumption is often used in many electronic designs.

QUESTIONS

A. Calculate the current through the silicon diode, as shown in Figure 2.18.

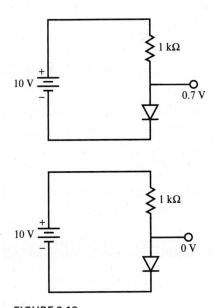

FIGURE 2.18

$V_D =$ _____

$V_R = V_S - V_D =$ _____

$I = \dfrac{V_R}{R} =$ _____

$I_D =$ _____

B. Calculate the current through the perfect diode, as shown in Figure 2.18.

$V_D =$ _____

$V_R = V_S - V_D =$ _____

$I = \dfrac{V_R}{R} =$ _____

$I_D =$ _____

ANSWERS

A. 0.7 volt; 9.3 volt; 9.3 mA; 9.3 mA

B. 0 volt; 10 volt; 10 mA; 10 mA

---

14 The difference in the values of the two currents found in problem 13 is less than 10 percent of the total current. That is, 0.7 mA is less than 10 percent of 10 mA. Many electronic components have a plus or minus 5 percent tolerance in their nominal values. This means that a 1 kΩ resistor can be anywhere from 950 ohms to 1,050 ohms, meaning that the value of current through a resistor can vary plus or minus 5 percent.

Because of a slight variance in component values, calculations are often simplified if the simplification does not change values by more than 10 percent. Therefore, a diode is often assumed to be perfect when the supply voltage is 10 volts or more.

QUESTIONS

A. Examine the circuit in Figure 2.19. Is it safe to assume that the diode is perfect?

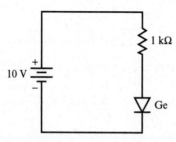

FIGURE 2.19

B. Calculate the current through the diode.

ANSWERS

A. Yes, it can be considered a perfect diode.

B. $I = 10\,mA$

---

15 When a current flows through a diode, it causes heating and power dissipation, just as with a resistor. The power formula for resistors is $P = V \times I$. This same formula can be applied to diodes to find the power dissipation.

To calculate the power dissipation in a diode, you must first calculate the current as shown previously. The voltage drop in this formula is assumed to be 0.7 volt for a silicon diode, even if you considered it to be 0 volts when calculating the current.

For example, a silicon diode has 100 mA flowing through it. Determine how much power the diode dissipates.

$$P = (0.7\,volt)(100\,mA) = 70\,mW$$

QUESTION

Assume a current of 2 amperes is flowing through a silicon diode. How much power is being dissipated?_____

_____

ANSWER

$$P = (0.7\,volts)(2\,amperes) = 1.4\,watts$$

16 Diodes are made to dissipate a certain amount of power, and this is quoted as a maximum power rating in the manufacturer's specifications of the diode.

Assume a silicon diode has a maximum power rating of 2 watts. How much current can it safely pass?

$$P = V \times I$$

$$I = \frac{P}{V}$$

$$= \frac{2\,watts}{0.7\,volt}$$

2.86 amperes (rounded off to two decimal places)

Provided the current in the circuit does not exceed this, the diode cannot overheat and burn out.

QUESTION

Suppose the maximum power rating of a germanium diode is 3 watts. What is its highest safe current?_____

ANSWER

$$I = \frac{3\,\text{watts}}{0.3\,\text{volt}} = 10\,\text{amperes}$$

17 Answer the following questions for another example.

QUESTIONS

A. Could a 3-watt silicon diode carry the current calculated for the germanium diode for problem 16?_____

_____

B. What would be its safe current?_____

_____

ANSWERS

A. No, 10 amperes would cause a power dissipation of 7 watts, which would burn up the diode.

B. $I = \dfrac{3}{0.7} = 4.3\,\text{amperes}$

Any current less than this would be safe.

18 The next several examples concentrate on finding the current through the diode. Look at the circuit shown in Figure 2.20.

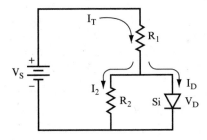

FIGURE 2.20

The total current from the battery flows through R_1, and then splits into I_2 and I_D. I_2 flows through R_2, and I_D flows through the diode.

QUESTIONS

A. What is the relationship between I_T, I_2, and I_D? _____

_____

B. What is the value of V_D? _____

_____

ANSWERS

A. Remember KCL, $I_T = I_2 + I_D$

B. $V_D = 0.7$ volt

19 To find I_D, you need to go through the following steps because there is no way to find I_D directly:

1. Find I_2. This is done using $V_D = R_2 \times I_2$.
2. Find V_R. For this, use $V_R = V_S - V_D$ (KVL again).
3. Find I_T (the current through R_1). Use $V_R = I_T \times R_1$.
4. Find I_D. This is found by using $I_T = I_2 + I_D$ (KCL again).

To find I_D in the circuit shown in Figure 2.21, go through these steps, and then check your answers.

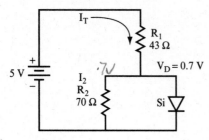

FIGURE 2.21

QUESTIONS

A. $I_2 = $ _____

B. $V_R = $ _____

C. $I_T = $ _____

D. $I_D = $ _____

ANSWERS

A. $I_2 = \dfrac{V_D}{R_2} = \dfrac{0.7\,\text{volt}}{R_2} = \dfrac{0.7\,\text{volt}}{70\,\text{ohms}} = 0.01\,\text{ampere} = 10\,\text{mA}$

B. $V_R = V_S - V_D = 5\,\text{volts} - 0.7\,\text{volt} = 4.3\,\text{volts}$

C. $I_T = \dfrac{V_R}{R_1} = \dfrac{4.3\,\text{volts}}{43\,\text{ohms}} = 0.1\,\text{ampere} = 100\,\text{mA}$

D. $I_D = I_T - I_2 = 100\,\text{mA} - 10\,\text{mA} = 90\,\text{mA}$

20 For this problem, refer to your answers in problem 19.

QUESTION

What is the power dissipation of the diode in problem 19? _____

ANSWER

$P = V_D \times I_D = (0.7 \text{ volt})(90 \text{ mA}) = 63 \text{ milliwatts}$

21 To find the current in the diode for the circuit shown in Figure 2.22, answer the following questions in order.

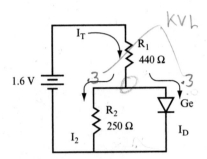

FIGURE 2.22

QUESTIONS

A. $I_2 =$ _____

B. $V_R =$ _____ $1.6 - .3 = 1.3v$ _____

C. $I_T =$ _____ $1.3v \div 440 = 3ma$ _____

D. $I_D =$ _____ $I_T - I_2 : \quad 3ma - 1.2ma = 1.8ma$

ANSWERS

A. $I_2 = \dfrac{0.3}{250} = 1.2 \text{ mA}$

B. $V_R = V_S - V_D = 1.6 - 0.3 = 1.3 \text{ volts}$

C. $I_T = \dfrac{V_R}{R_1} = \dfrac{1.3}{440} = 3 \text{ mA}$

D. $I_D = I_T - I_2 = 1.8 \text{ mA}$

If you want to take a break soon, this is a good stopping point.

DIODE BREAKDOWN

22 Earlier, you read that if the circuit in Project 2.1 was not working correctly, then the diode may be in backward. If you place the diode in the circuit backward—as shown on the right in Figure 2.23—then almost no current flows. In fact, the current flow is so small, it can be said that no current flows. The V-I curve for a reversed diode looks like the one shown on the left in Figure 2.23.

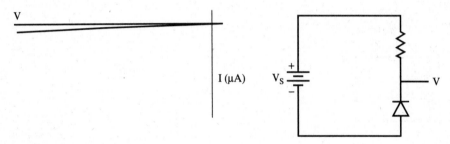

FIGURE 2.23

The V-I curve for a perfect diode would show zero current for all voltage values. But for a real diode, a voltage is reached where the diode "breaks down" and the diode allows a large current to flow. The V-I curve for the diode breakdown would then look like the one in Figure 2.24.

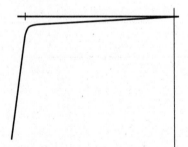

FIGURE 2.24

If this condition continues, the diode will burn out. You can avoid burning out the diode, even though it is at the breakdown voltage, by limiting the current with a resistor.

QUESTION

The diode in the circuit shown in Figure 2-25 will break down at 100 volts, and it can safely pass 20 mA without overheating at that voltage. Find the resistance in this circuit that would limit the current to 20 mA.

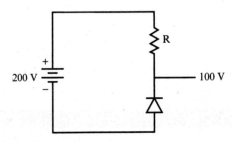

FIGURE 2.25

ANSWER

$V_R = V_S - V_D = 200$ volts $-$ 100 volts $=$ 100 volts

Because 1 ampere of current is flowing, then

$$R = \frac{V_R}{I} = \frac{100 \text{ volts}}{1 \text{ ampere}} = 100 \text{ ohms}$$

23 All diodes break down when connected in the reverse direction if excess voltage is applied to them. The breakdown voltage (which is a function of how the diode is made) varies from one type of diode to another. This voltage is quoted in the manufacturer's data sheet.

Breakdown is not a catastrophic process and does not destroy the diode. If the excessive supply voltage is removed, the diode can recover and operate normally. You can use it safely many more times, provided the current is limited to prevent the diode from burning out.

A diode always breaks down at the same voltage, no matter how many times it is used.

The breakdown voltage is often called the *peak inverse voltage (PIV)* or the *peak reverse voltage (PRV)*. Following are the PIVs of some common diodes:

Diode	PIV
1N4001	50 volts
1N4002	100 volts
1N4003	200 volts
1N4004	400 volts
1N4005	600 volts
1N4006	800 volts

QUESTIONS

A. Which can permanently destroy a diode, excessive current or excessive voltage?

B. Which is more harmful to a diode, breakdown or burnout?

ANSWERS

A. Excessive current. Excessive voltage cannot harm the diode if the current is limited.

B. Burnout. Breakdown is not necessarily harmful, especially if the current is limited.

INSIDE THE DIODE

At the junction of the N and P type regions of a diode, electrons from the N region are trapped by holes in the P region, forming a depletion region as illustrated in the following figure.

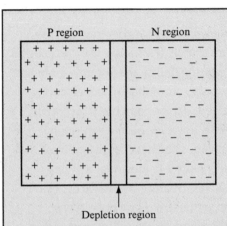

Depletion region

When the electrons are trapped by holes, they can no longer move, which is what produces this depletion region that contains no mobile electrons. The static electrons give this region the properties of an electrical insulator.

You apply a forward-biased voltage to a diode by applying negative voltage to the N region and positive voltage to the P region. Electrons in the N region are repelled by the negative voltage, pushing more electrons into the depletion region. However, the electrons in the N region are also repelled by the electrons already in the depletion region. (Remember that like charges repel each other.) When the forward-biased voltage is sufficiently high (0.7 volts for silicon diodes, and 0.3 volts for germanium diodes), the depletion region is eliminated. As the voltage is raised further, the negative voltage, in repelling electrons in the N region, pushes them into the P region, where they are attracted by the positive voltage.

This combination of repulsive and attractive forces allows current to flow, as illustrated in the following figure.

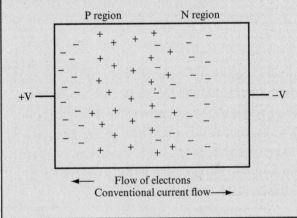

Continued

(continued)

Note that an interesting aspect of diodes is that while the N and P regions of the diode have mobile charges, they do not have a net charge. The mobile charges (electrons in N regions and holes in P regions) are simply donated by the impurity atoms used to dope the semiconductor material. The impurity atoms have more electrons in the N region and less in the P region than are needed to bond to the adjacent silicon or germanium atoms in the crystalline structure. However, because the atoms have the same number of positive and negative charges, the net charge is neutral.

In a similar way, metal is a conductor because it has more electrons than are needed to bond together the atoms in its crystalline structure. These excess electrons are free to move when a voltage is applied, either in a metal or an N type semiconductor, which makes the material electrically conductive. Electrons moving through the P region of the diode jump between the holes, which physicists model as the holes moving. In a depletion region, where electrons have been trapped by holes, there is a net negative charge.

You apply a reverse-biased voltage to a diode by applying positive voltage to the N type region and negative voltage to the P type region. In this scenario, electrons are attracted to the positive voltage, which pulls them away from the depletion region. No current can flow unless the voltage exceeds the reverse breakdown voltage.

Diodes are marked with one band indicating the cathode (the N region) end of the diode. The band on the diode corresponds to the bar on the diode symbol, as shown in the following figure. Use this marking to orient diodes in a circuit.

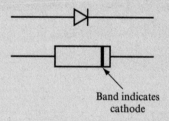

Band indicates
cathode

The part number is marked on diodes. However, the part numbers don't tell you much about the diode. For example, the 1N4001 is a silicon diode that can handle a peak reverse voltage of 50 volts, whereas the 1N270 is a germanium diode that can handle a peak reverse voltage of 80 volts. Your best bet is to refer to the manufacturer's data sheet for the diode peak reverse voltage and other characteristics. You can easily look up data sheets on the Internet. Also, you can find links to the data sheets for the components used in this book on the website at
`www.buildinggadgets.com/index_datasheets.htm`.

THE ZENER DIODE

24 Diodes can be manufactured so that breakdown occurs at lower and more precise voltages than those just discussed. These types of diodes are called *zener diodes*, so named because they exhibit the "Zener effect"—a particular form of voltage breakdown. At the *zener voltage*, a small current flows through the zener diode. This current must be maintained to keep the diode at the *zener point*. In most cases, a few milliamperes are all that is required. Figure 2.26 shows the zener diode symbol and a simple circuit.

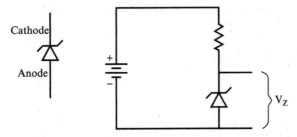

FIGURE 2.26

In this circuit, the battery determines the applied voltage. The zener diode determines the voltage drop (labeled V_z) across it. The resistor determines the current flow. Zeners are used to maintain a constant voltage at some point in a circuit.

QUESTION

Why are zeners used for this purpose, rather than ordinary diodes?_____

_____

ANSWER

Because zeners have a precise breakdown voltage.

---

25 Examine an application in which a constant voltage is wanted—for example, a lamp driven by a DC generator. In this example, when the generator turns at full speed, it puts out 50 volts. When it runs more slowly, the voltage can drop to 35 volts. You want to illuminate a 20-volt lamp with this generator. Assume that the lamp draws 1.5 amperes. Figure 2.27 shows the circuit.

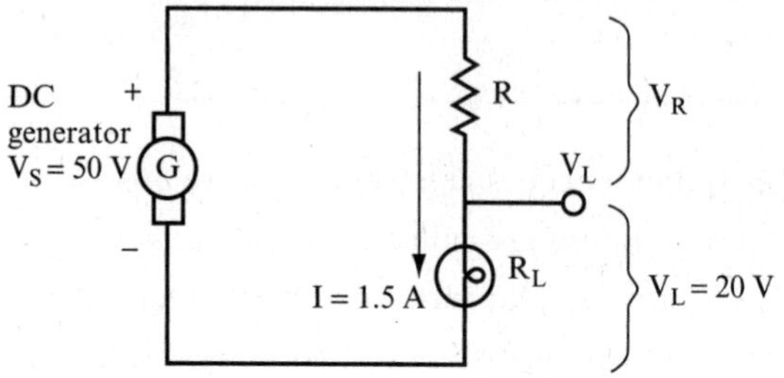

FIGURE 2.27

You need to determine a suitable value for the resistance. Follow these steps to find a suitable resistance value:

1. Find R_L, the lamp resistance. Use the following formula:

   $$R_L = \frac{V_L}{I}$$

2. Find V_R. Use $V_S = V_R + V_L$.

3. Find R. Use the following formula:

   $$R = \frac{V_R}{I}$$

QUESTIONS

Work through these steps, and write your answers here.

A. $R_L = $ _____

B. $V_R = $ _____

C. $R = $ _____

ANSWERS

A. $R_L = \dfrac{20 \text{ volts}}{1.5 \text{ amperes}} = 13.33$ ohms

B. $V_R = 50$ volts $- 20$ volts $= 30$ volts

C. $R = \dfrac{50 \text{ volts} - 20 \text{ volts}}{1.5 \text{ amperes}} = \dfrac{30 \text{ volts}}{1.5 \text{ amperes}} = 20$ ohms

26 Assume now that the 20-ohm resistor calculated in problem 25 is in place, and the voltage output of the generator drops to 35 volts, as shown in Figure 2.28. This is similar to what happens when a battery gets old. Its voltage level decays and it will no longer have sufficient voltage to produce the proper current. This results in the lamp glowing less brightly, or perhaps not at all. Note, however, that the resistance of the lamp stays the same.

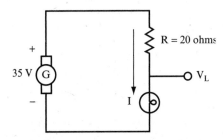

FIGURE 2.28

QUESTIONS

A. Find the total current flowing. Use the following formula:

$$I_T = \frac{V_S}{(R + R_L)}$$

$I_T =$ _____

B. Find the voltage drop across the lamp. Use $V_L = I_T \times R_L$.

$V_L =$ _____

C. Have the voltage and current increased or decreased?

_____

ANSWERS

A. $I_T = \dfrac{35 \text{ volts}}{20\,\Omega + 13.3\,\Omega} = \dfrac{35 \text{ volts}}{33.3\,\Omega} = 1.05 \text{ amperes}$

B. $V_L = 1.05 \text{ amperes} \times 13.3\,\Omega = 14 \text{ volts}$

C. Both have reduced in value.

27 In many applications, a lowering of voltage across the lamp (or some other compo- nent) may be unacceptable. You can prevent this by using a zener diode, as shown in the circuit in Figure 2.29.

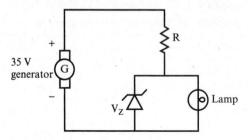

FIGURE 2.29

If you choose a 20-volt zener (that is, one that has a 20-volt drop across it), then the lamp always has 20 volts across it, no matter what the output voltage is from the genera- tor (provided, of course, that the output from the generator is always above 20 volts).

QUESTIONS

Say that the voltage across the lamp is constant, and the generator output drops.

A. What happens to the current through the lamp? _____

_____

B. What happens to the current through the zener? _____

_____

ANSWERS

A. The current stays constant because the voltage across the lamp stays constant.

B. The current decreases because the total current decreases.

28 To make this circuit work and keep 20 volts across the lamp at all times, you must find a suitable value of R. This value should allow sufficient total current to flow to pro- vide 1.5 amperes required by the lamp, and the small amount required to keep the diode at its zener voltage. To do this, you start at the "worst case" condition. ("Worst case" design is a common practice in electronics. It is used to ensure that equipment can work under the most adverse conditions.) The worst case here would occur when the generator puts out only 35 volts. Figure 2.30 shows the paths that current would take in this circuit.

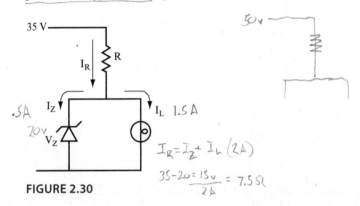

35 V

I_R R

.5A I_Z I_L 1.5 A

70v V_Z

50v

$I_R = I_Z + I_L$ (2A)

$\dfrac{35 - 20 = 15v}{2A} = 7.5\,\Omega$

FIGURE 2.30

Find the value of R that enables 1.5 amperes to flow through the lamp. How much current can flow through the zener diode? You can choose any current you like, provided it is above a few milliamperes, and provided it does not cause the zener diode to burn out. In this example, assume that the zener current I_Z is 0.5 amperes.

QUESTIONS

A. What is the total current through R?

$I_R =$ _____

B. Calculate the value of R.

$R =$ _____

ANSWERS

A. $I_R = I_L + I_Z = 1.5$ amperes $+ 0.5$ amperes $= 2$ amperes

B. $R = \left(\dfrac{V_S - V_Z}{I_R}\right) = \dfrac{(35 \text{ volts} - 20 \text{ volts})}{2 \text{ amperes}} = 7.5 \text{ ohms}$

A different choice of I_Z here would produce another value of R.

29 Now, take a look at what happens when the generator supplies 50 volts, as shown in Figure 2.31.

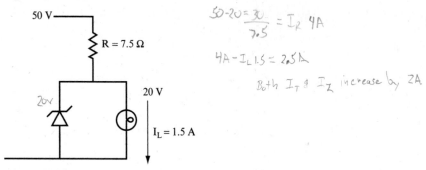

Handwritten annotations:
$$\frac{50-20=30}{7.5} = I_R \ 4A$$

$$4A - I_L \ 1.5 = 2.5A$$

Both I_T & I_Z increase by 2A

FIGURE 2.31

Because the lamp still has 20 volts across it, it can still draw only 1.5 amperes. But the total current and the zener current change.

QUESTIONS

A. Find the total current through R.

$I_R = $ _____

B. Find the zener current.

$I_Z = $ _____

ANSWERS

A. $I_R = \dfrac{(V_S - V_Z)}{R} = \dfrac{(50-20)}{7.5} = 4\,\text{amperes}$

B. $I_Z = I_R - I_L = 4 - 1.5 = 2.5\ \text{amperes}$

30 Although the lamp voltage and current remain the same, the total current and the zener current both changed.

QUESTIONS

A. What has happened to I_T (I_R)? _____

_____

B. What has happened to I_Z? _____

_____

THE ZENER DIODE 81

ANSWERS

A. I_T has increased by 2 amperes.

B. I_Z has increased by 2 amperes.

The increase in I_T flows through the zener diode and not through the lamp.

31 The power dissipated by the zener diode changes as the generator voltage changes.

QUESTIONS

A. Find the power dissipated when the generator voltage is 35 volts. _____0 watts_____

B. Now, find the power when the generator is at 50 volts. _____50 watts_____

ANSWERS

A. $P_Z = V \times I = (20 \text{ volts}) (0.5 \text{ ampere}) = 10 \text{ watts}$

B. $P_Z = V \times I = (20 \text{ volts}) (2.5 \text{ ampere}) = 50 \text{ watts}$

If you use a zener diode with a power rating of 50 watts or more, it does not burn out.

32 Use Figure 2.32 to answer the following question.

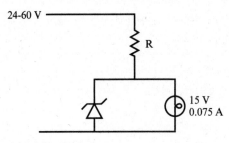

FIGURE 2.32

QUESTION

For the circuit shown in Figure 2.32, what power rating should the zener diode have? The current and voltage ratings of the lamp are given. _____

_____

ANSWER

At 24 volts, assuming a zener current of 0.5 ampere:

$$R = \frac{9}{0.575} = 15.7 \text{ ohms}$$

At 60 volts:

$$I_R = \frac{45}{15.7} = 2.87 \text{ amperes; therefore } I_Z \approx 2.8 \text{ amperes}$$

$$P_Z = (15 \text{ volts})(2.8 \text{ amperes}) = 42 \text{ watts}$$

PROJECT 2.2: The Zener Diode Voltage Regulator

OBJECTIVE

The objective of this project is to measure the voltage applied to the lamp, and the current through the lamp for different supply voltages, demonstrating the use of a zener diode to provide a steady voltage and current to a lamp when the supply voltage changes.

GENERAL INSTRUCTIONS

While the circuit is set up, measure the lamp current, zener diode current, and supply voltage as the voltage from the 9-volt battery drops.

Parts List

❑ One 9-volt battery

❑ One battery snap connector

❑ Two multimeters set to measure current (mA)

❑ One multimeter set to measure DC voltage

- ❏ One 56-ohm, 0.5-watt resistor
- ❏ One 1N4735A zener diode
- ❏ One breadboard
- ❏ One lamp rated for approximately 25 mA at 6 volts. (Part # 272-1140 from Radio Shack is a good fit for this project.)
- ❏ Two terminal blocks

STEP-BY-STEP INSTRUCTIONS

Set up the circuit shown on Figure 2.33. The circled "A" designates a multimeter set to measure current, and the circled "V" designates a multimeter set to measure DC voltage. If you have some experience in building circuits, this schematic (along with the previous parts list) should provide all the information you need to build the circuit. If you need a bit more help in building the circuit, look at the photos of the completed circuit in the "Expected Results" section.

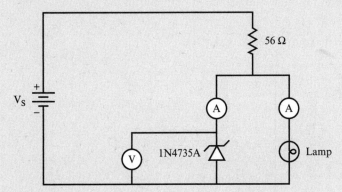

FIGURE 2.33

Carefully check your circuit against the diagram, especially the direction of the battery and the diode. The diode has a band at one end. Connect the lead at the end of the diode without the band to the ground bus on the breadboard.

After you check your circuit, follow these steps, and record your measurements in the blank table following the steps:

1. Measure and record the supply voltage.

2. Measure and record the lamp current.

3. Measure and record the zener current.

4. Wait 30 minutes.

5. Measure and record the new values of voltage and current.

6. Repeat steps 4 and 5 four times.

Time (Minutes)	V_S (Volts)	I_L (mA)	I_Z (mA)
0			
30			
60 (1 hr)			
90			
120 (2 hr)			

EXPECTED RESULTS

Figure 2.34 shows the breadboarded circuit for this project.

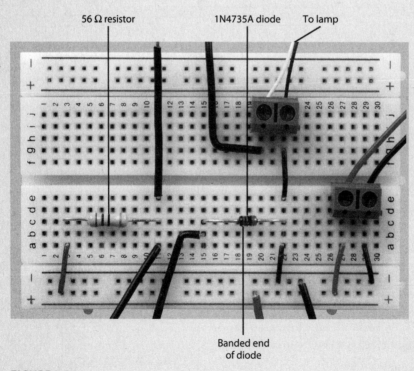

FIGURE 2.34

Figure 2.35 shows the test setup for this project.

To multimeter set to mA

To multimeter set to mA

To multimeter set to volts

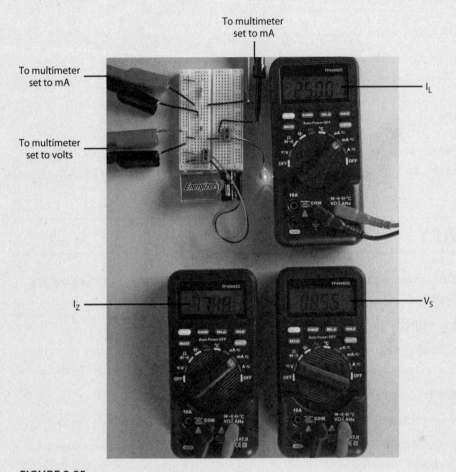

I_L

I_Z

V_S

FIGURE 2.35

Compare your measurements with the ones shown in the following table. You should see a similar trend in the measured values, but not exactly the same values.

Time (Minutes)	V_S (Volts)	I_L (mA)	I_Z (mA)
0	9.06	25.1	25.6
30	8.45	25.0	15.6
60 (1 hr)	8.21	25.0	11.8
90	8.03	24.9	9.0

Continued

(continued)

Time (Minutes)	V_S (Volts)	I_L (mA)	I_Z (mA)
120 (2 hr)	7.91	24.9	7.0
150	7.82	24.9	5.7
180 (3 hr)	7.76	24.9	4.6
210	7.70	24.9	3.7
240 (4 hrs)	7.65	24.9	2.9

As you can see in this data, even though the supply voltage dropped by approximately 15 percent, the lamp current stayed roughly constant, showing less than a 1 percent drop.

SUMMARY

Semiconductor diodes are used extensively in modern electronic circuits. Following are the main advantages of semiconductor diodes:

- They are small.
- They are rugged and reliable if properly used. You must remember that excessive reverse voltage or excessive forward current could damage or destroy the diode.
- Diodes are easy to use because there are only two connections to make.
- They are inexpensive.
- They can be used in all types of electronic circuits, from simple DC controls to radio and TV circuits.
- They can be made to handle a wide range of voltage and power requirements.
- Specialized diodes (which have not been covered here) can perform particular functions, which no other components can handle.
- Finally, as you see in Chapter 3, "Introduction to the Transistor," diodes are an integral part of transistors.

All the many uses of semiconductor diodes are based on the fact they conduct in *one direction only*. Diodes are often used for the following:

- Protecting circuit components from voltage spikes
- Converting AC to DC
- Protecting sensitive components from high-voltage spikes

- Building high-speed switches
- Rectifying radio frequency signals

SELF-TEST

The following questions test your understanding of this chapter. Use a separate sheet of paper for your diagrams or calculations. Compare your answers with the answers that follow the test.

1. Draw the circuit symbol for a diode, labeling each terminal. _Anode ▷|◁ Cathode_ _____

   _____

2. What semiconductor materials are used in the manufacture of diodes? _Germ /Silic._

   _____

3. Draw a circuit with a battery, a resistor, and a forward-biased diode. _____

   _____

4. What is the current through a reverse-biased perfect diode? _O_ _____

   _____

5. Draw a typical V-I curve of a forward-biased diode. Show the knee voltage. _⌐_ _____

   _____

6. What is the knee voltage for silicon? _.7 v_ _____

   _____

   Germanium? _.3 v_ _____

   _____

7. In the circuit shown in Figure 2.36, V_s = 10 volts and R = 100 ohms. Find the current through the diode, assuming a perfect diode. _.10 A_ _____

   _____

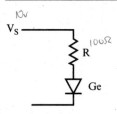

FIGURE 2.36

8. Calculate question 7 using these values: V_s = 3 volts and R = 1 kΩ. _____3 ma_____

_____

9. In the circuit shown in Figure 2.37, find the current through the diode.

V_s = 10 volts

R_1 = 10 kΩ

R_2 = 1 kΩ

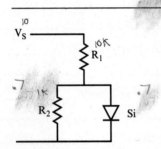

1) .7 ma

2 $9.3/_{10k}$ = .9 ma

3 .9 ma - .7 ma = .2 ma

FIGURE 2.37

10. In the circuit shown in Figure 2.38, find the current through the zener diode.

V_s = 20 volts

V_z = 10 volts

R_1 = 1 kΩ

R_2 = 2 kΩ

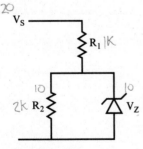

1) $10/_{2k}$ = 5 ma

2) 20-10 = $10/_{1K}$ = 10 ma

3) 10 - 5 = 5 ma

.005 x 10 = .05 watt

1)

2) 45-10 = $35/_{1K}$ = 35 m

3) 35-5 = 30 ma

.03 x 10 = .3 watt

FIGURE 2.38

11. If the supply voltage for question 10 increases to 45 volts, what is the current in the zener diode? _____

_____

12. What is the maximum power dissipated for the diode in questions 10 and 11?_____

_____

ANSWERS TO SELF-TEST

If your answers do not agree with those given here, review the problems indicated in parentheses before you go to Chapter 3, "Introduction to the Transistor."

1.	See Figure 2.39. 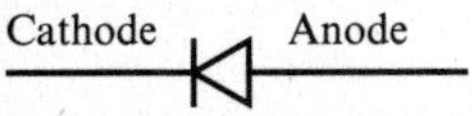 Cathode ——▷|—— Anode **FIGURE 2.39**	(problem 3)
2.	Germanium and silicon.	(problem 1)
3.	See Figure 2.40. 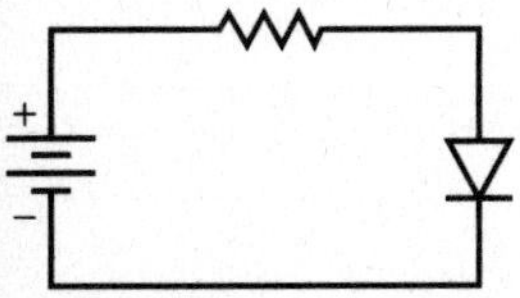 **FIGURE 2.40**	(problem 4)
4.	There is zero current flowing through the diode.	(problem 6)
5.	See Figure 2.41. 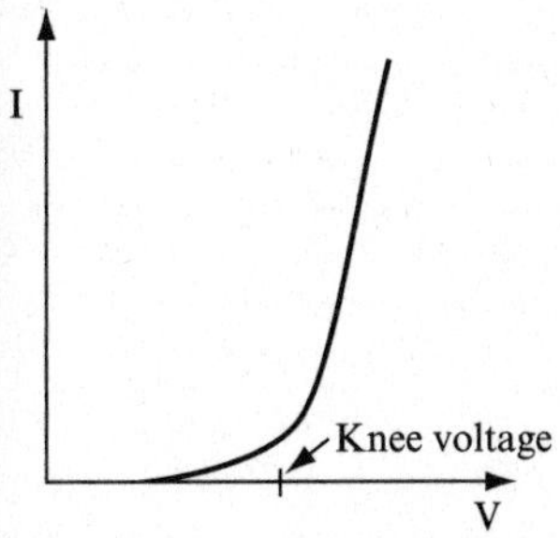 Knee voltage **FIGURE 2.41**	(Project 2.1)

Continued

(continued)

6.	Si = 0.7 volt; Ge = 0.3 volt (These are approximate.)	(Project 2.1)
7.	I_D = 100 mA.	(problem 12)
8.	As V_S = 3 volt, do not ignore the voltage drop across the diode. Thus, I_D = 2.7 mA.	(problem 12)
9.	Ignore V_D in this case. Thus, I_D = 0.3 mA. If V_D is not ignored, I_D = 0.23 mA.	(problem 19)
10.	I_Z = 5 mA.	(problem 29)
11.	I_Z = 30 mA.	(problem 29)
12.	The maximum power will be dissipated when I_Z is at its peak value of 30 mA. Therefore, $P_Z(MAX)$ = 0.30 watt.	(problem 31)

Introduction to the Transistor

The transistor is undoubtedly the most important modern electronic component because it has enabled great and profound changes in electronics and in your daily lives since its discovery in 1948.

This chapter introduces the transistor as an electronic component that acts similarly to a simple mechanical switch, and it is actually used as a switch in many modern electronic devices. A transistor can be made to conduct or not conduct an electric current—exactly what a mechanical switch does.

Most transistors used in electronic circuits are *bipolar junction transistors* (*BJTs*), commonly referred to as *bipolar transistors junction field effect transistors* (*JFETs*) or *metal oxide silicon field effect transistors* (*MOSFETs*). This chapter (along with Chapter 4,

"The Transistor Switch," and Chapter 8, "Transistor Amplifiers") illustrates how BJTs and JFETs function and how they are used in electronic circuits. Because JFETs and MOSFETs function in similar fashion, MOSFETs are not covered here.

Projects in this chapter can help you to build a simple one-transistor circuit. You can easily set up this circuit on a home workbench. You should take the time to obtain the few components required, and actually build and operate the circuit.

In Chapter 4, you continue to study transistor circuits and the operation of the transistor as a switch. In Chapter 8, you learn how a transistor can be made to operate as an amplifier. In this mode, the transistor produces an output that is a magnified version of an input signal, which is useful because many electronic signals require amplification. These chapters taken together present an easy introduction to how transistors function, and how they are used in basic electronic circuits.

When you complete this chapter, you can do the following:

- Describe the basic construction of a BJT.

- Describe the basic construction of a JFET.

- Specify the relationship between base and collector current in a BJT.

- Specify the relationship between gate voltage and drain current in a JFET.

- Calculate the current gain for a BJT.

- Compare the transistor to a simple mechanical switch.

UNDERSTANDING TRANSISTORS

1 The diagrams in Figure 3.1 show some common transistor cases (also called *packages*). The cases protect the semiconductor chip on which the transistor is built and provide leads that can be used to connect it to other components. For each transistor, the diagrams show the lead designations and how to identify them based on the package design.

Transistors can be soldered directly into a circuit, inserted into sockets, or inserted into breadboards. When soldering, you must take great care because transistors can be destroyed if overheated. A heat sink clipped to the transistor leads between the solder joint and the transistor case can reduce the possibility of overheating. If you use a socket, you can avoid exposing the transistor to heat by soldering the connections to the socket before inserting the transistor.

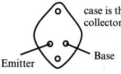

FIGURE 3.1

QUESTIONS

A. How many leads are there on most transistors? _____

_____

B. Where there are only two leads, what takes the place of the third lead? _____

_____

C. What are the three leads or connections called? _____

_____

D. Why should you take care when soldering transistors into a circuit? _____

_____

ANSWERS

A. Three.

B. The case can be used instead, as indicated in the diagram on the right side of Figure 3.1. (This type of case is used for power transistors.)

C. Emitter, base, and collector.

D. Excessive heat can damage a transistor.

2 You can think of a bipolar junction transistor as functioning like two diodes, connected back-to-back, as illustrated in Figure 3.2.

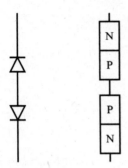

FIGURE 3.2

However, in the construction process, one important modification is made. Instead of two separate P regions, as shown in Figure 3.2, only one thin region is used, as shown in Figure 3.3.

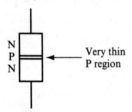

FIGURE 3.3

QUESTION

Which has the thicker P region, the transistor shown in Figure 3.3 or two diodes connected back-to-back? _____

_____

ANSWER

Two diodes. The transistor has a thin P region.

3 Because two separate diodes wired back-to-back share two thick P regions, they cannot behave like a transistor.

QUESTION

Why don't two diodes connected back-to-back function like a transistor? _____

_____

ANSWER

The transistor has one thin P region, whereas the diodes share two thick P regions.

4 The three terminals of a transistor (the base, the emitter, and the collector) connect, as shown in Figure 3.4.

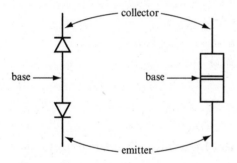

FIGURE 3.4

When talking about a transistor as two diodes, you refer to the diodes as the *base-emitter diode* and the *base-collector diode*.

Figure 3.5 shows the symbol used in circuit diagrams for the transistor, with the two diodes and the junctions shown for comparison. Because of the way the semiconductor materials are arranged, this is known as an *NPN transistor*.

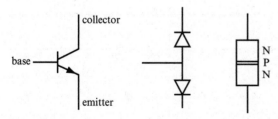

FIGURE 3.5

QUESTION

Which transistor terminal includes an arrowhead? _____

_____

ANSWER

The emitter

5 It is also possible to make transistors with a PNP configuration, as shown in Figure 3.6.

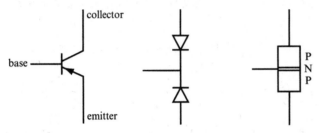

FIGURE 3.6

Both NPN and PNP type transistors can be made from either silicon or germanium.

QUESTIONS

A. Draw a circuit symbol for both an NPN and a PNP transistor. (Use a separate sheet of paper for your drawings.)

B. Which of the transistors represented by these symbols might be silicon? _____

_____

C. Are silicon and germanium ever combined in a transistor? _____

_____

ANSWERS

A. See Figure 3.7.

B. Either or both could be silicon. (Either or both could also be germanium.)

C. Silicon and germanium are *not* mixed in any commercially available transistors. However, researchers are attempting to develop ultra-fast transistors that contain both silicon and germanium.

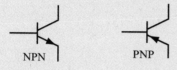

NPN   PNP

FIGURE 3.7

6 Take a look at the simple examples using NPN transistors in this and the next few problems.

If a battery is connected to an NPN transistor, as shown in Figure 3.8, then a current will flow in the direction shown.

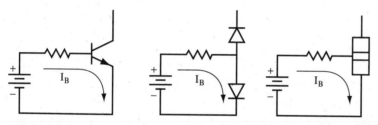

FIGURE 3.8

The current flowing through the base-emitter diode is called *base current* and is represented by the symbol I_B.

QUESTION

Would base current flow if the battery were reversed? Give a reason for your answer.

ANSWER

Base current would not flow because the diode would be back-biased.

7 In the circuit shown in Figure 3.9, you can calculate the base current using the techniques covered in Chapter 2, "The Diode."

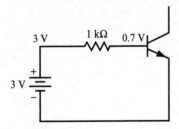

FIGURE 3.9

QUESTION

Find the base current in the circuit shown in Figure 3.9. (*Hint:* Do not ignore the 0.7-volt drop across the base-emitter diode.)

$$I_B = \underline{\hspace{6cm}}$$

ANSWER

Your calculations should look something like this:

$$I_B = \frac{(V_S - 0.7\,\text{volt})}{R} = \frac{(3 - 0.7)}{1\,k\Omega} = \frac{2.3\,\text{volts}}{1\,k\Omega} = 2.3\,\text{mA}$$

8 For the circuit shown in Figure 3.10, because the 10 volts supplied by the battery is much greater than the 0.7-volt diode drop, you can consider the base-emitter diode to be a perfect diode and thus assume the voltage drop is 0 volts.

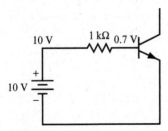

FIGURE 3.10

QUESTION

Calculate the base current.

$I_B =$ _____

ANSWER

$$I_B = \frac{(10-0)}{1\,k\Omega} = \frac{10}{1\,k\Omega} = 10\,mA$$

9 Look at the circuit shown in Figure 3.11.

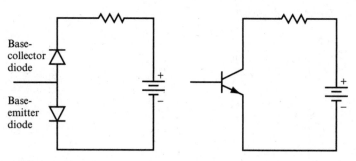

FIGURE 3.11

QUESTION

Will current flow in this circuit? Why or why not? _____

_____

ANSWER

It cannot flow because the base-collector diode is reverse-biased.

10 Examine the circuit shown in Figure 3.12. Batteries are connected to both the base and collector portions of the circuit.

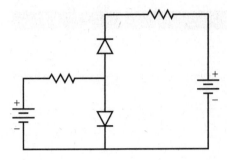

FIGURE 3.12

When you connect batteries to both the base and the collector portions of the circuit, currents flowing through the circuit demonstrate a key characteristic of the transistor. This characteristic is sometimes called *transistor action*—if base current flows in a transistor, collector current will also flow.

Examine the current paths shown in Figure 3.13.

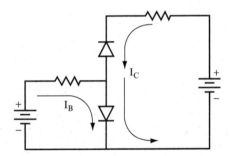

FIGURE 3.13

QUESTIONS

A. What current flows through the base-collector diode? _____

_____

B. What current flows through the base-emitter diode? _____

_____

C. Which of these currents causes the other to flow? _____

_____

ANSWERS

A. I_C (the collector current).

B. I_B and I_C. Both of them flow through the base-emitter diode.

C. Base current causes collector current to flow.

No current flows along the path shown by the dotted line in Figure 3.14 from the collector to the base.

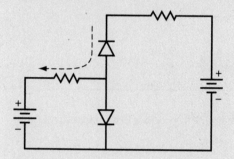

FIGURE 3.14

11 Up to now, you have studied the NPN bipolar transistor. PNP bipolar transistors work in the same way as NPN bipolar transistors and can also be used in these circuits.

There is, however, one important circuit difference, which is illustrated in Figure 3.15. The PNP transistor is made with the diodes oriented in the reverse direction from the NPN transistor.

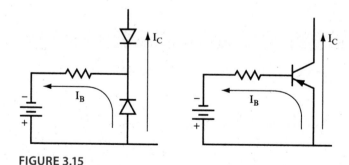

FIGURE 3.15

QUESTIONS

Compare Figure 3.15 with Figure 3.13. How are the circuits different relative to the following?

A. Battery connections: _____

B. Current flow: _____

ANSWERS

A. The battery is reversed in polarity.

B. The currents flow in the opposite direction.

12 Figure 3.16 shows the battery connections necessary to produce both base current and collector current in a circuit that uses a PNP transistor.

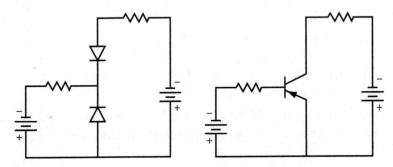

FIGURE 3.16

QUESTION

In which direction do these currents circulate—clockwise or counterclockwise? _____

_____

ANSWER

Base current flows counterclockwise.
Collector current flows clockwise.

As stated earlier, NPN and PNP bipolar transistors work in much the same way: Base current causes collector current to flow in both. The only significant difference in using a PNP versus an NPN bipolar transistor is that the polarity of the supply voltage (for both the base and collector sections of the circuit) is reversed. To avoid confusion, bipolar transistors used throughout the rest of this book are NPNs.

13 Consider the circuit shown in Figure 3.17. It uses only one battery to supply voltage to both the base and the collector portions of the circuit. The path of the base current is shown in the diagram.

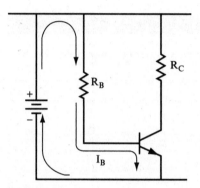

FIGURE 3.17

QUESTIONS

A. Name the components through which the base current flows. _____
_____

B. Into which terminal of the transistor does the base current flow? _____
_____

C. Out of which transistor terminal does the base current flow? _____
_____

D. Through which terminals of the transistor does base current not flow? _____
_____

ANSWERS

A. The battery, the resistor R_B, and the transistor

B. Base

C. Emitter

D. Collector

14 Take a moment to recall the key physical characteristic of the transistor.

QUESTION

When base current flows in the circuit shown in Figure 3.17, what other current can flow, and which components will it flow through? _____

_____

ANSWER

Collector current will flow through the resistor R_C and the transistor.

15 In Figure 3.18 the arrows indicate the path of the collector current through the circuit.

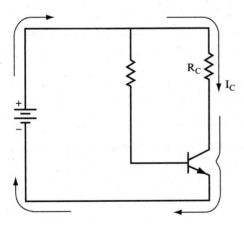

FIGURE 3.18

QUESTIONS

A. List the components through which the collector current flows. _____

_____

B. What causes the collector current to flow? _____

_____

ANSWERS

A. The resistor R_C, the transistor, and the battery.

B. Base current. (Collector current doesn't flow unless base current is flowing.)

16 The *ratio of the collector current to base current* in a transistor is called the *current gain*, which is represented by the symbol β, or beta. The collector current is always much larger than the base current. Typical values of β range from 10 to 300.

QUESTIONS

A. What is the ratio of collector current to base current called? _gain_____

_____

B. What is the symbol used to represent this? _β_____

_____

C. Which is larger—base or collector current? _Coll_____

_____

D. Look back at the circuit in problem 13. Will current be greater in R_B or in R_C? _R_c_

_____

ANSWERS

A. Current gain.

B. β.

C. Collector current is larger.

D. The current is greater in R_C because it is the collector current.

NOTE   The β introduced here is referred to in manufacturers' specification sheets as h_{FE}. Technically, it is referred to as the static or DC β. For the purposes of this chapter, it is called β. Discussions on transistor parameters in general, which are well covered in many textbooks, will not be covered here.

17 The mathematical formula for current gain is as follows:

$$\beta = \frac{I_C}{I_B}$$

In this equation, the following is true:

I_B = base current

I_C = collector current

The equation for β can be rearranged to give $I_C = \beta\, I_B$. From this, you can see that if no base current flows, no collector current flows. Also, if more base current flows, more collector current flows. This is why it's said that the base current controls the collector current.

QUESTION

Suppose the base current is 1 mA and the collector current is 150 mA. What is the current gain of the transistor? _____

_____

ANSWER

150

INSIDE THE BIPOLAR TRANSISTOR

Now take a closer look inside a bipolar transistor. In a bipolar transistor, with no voltages applied, two depletion regions exist. As shown in the following figure, one *depletion region* exists at the emitter and base junction, and one at the collector and base junction. There are no free electrons or holes in these depletion regions, which prevents any current from flowing.

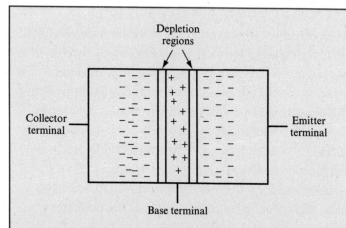

Depletion regions

Collector terminal

Emitter terminal

Base terminal

The following figure shows a bipolar transistor with a positive voltage applied to the base and a negative voltage applied to the emitter (forward-biasing the base-emitter diode), as well as a positive voltage applied to the collector. The forward bias on the base-emitter diode allows electrons to flow from the emitter into the base. A small fraction of these electrons would be captured by the holes in the base region and then flow out of the base terminal and through the base resistor. The base of a transistor is thin, which allows most of the electrons from the emitter to flow through the base and into the collector. The positive voltage on the collector terminal attracts these electrons, which flow out of the collector terminal and through the collector resistor.

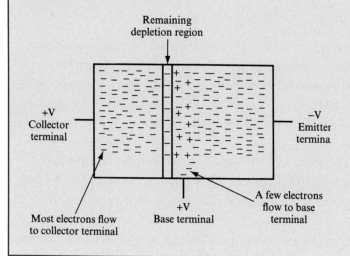

Remaining depletion region

+V
Collector terminal

−V
Emitter termina

Most electrons flow to collector terminal

+V
Base terminal

A few electrons flow to base terminal

Continued

continued

For a transistor with $\beta = 100$, only one electron flows out to the base terminal for every 100 electrons that flow to the collector terminal. β is controlled by two factors: the thickness of the base region and the relative concentration of the impurities providing holes in the P region to the concentration of the impurities providing electrons in the N regions. A thinner base region plus the lower relative concentration of holes allow more electrons to pass through the base without being captured, resulting in a higher β. (Remember, the direction that conventional electrical current flows in is opposite to the direction in which electrons flow.)

Before you connect any bipolar transistor to other components in a circuit, you must identify the emitter, base, and collector leads (referred to as the transistor's *pinout*) and determine whether the transistor is NPN or PNP.

Transistors are marked with a part number, such as 2N3904, 2N3906, BC337, and PN2222. However, the part numbers don't tell you much about the transistor. For these transistors, the 2N3904, BC337, and PN2222 are NPN, whereas the 2N3906 is a PNP, which is not obvious from the part number.

Also, the transistor pinout is not identified on the part number. For example, one NPN transistor, the BC337, uses the opposite leads for the emitter and collector than the 2N3904 transistor, as shown in the following figure.

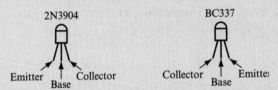

Your best bet is to refer to the manufacturer's data sheet for the transistor pinout and other characteristics. You can easily look up data sheets on the Internet. Also, you can find links to the data sheets for the transistors used in projects in this book on the website at `www.buildinggadgets.com/index_datasheets.htm`.

18 *Current gain* is a physical property of transistors. You can find the maximum and minimum values of β for a transistor part number in the manufacturers' published data sheets, however, you can determine the β of a particular transistor more accurately by experimenting.

One of the most frequently performed calculations in transistor work is to determine the values of either collector or base current, when β and the other current are known.

For example, suppose a transistor has 500 mA of collector current flowing, and you know it has a β value of 100. To find the base current, use the following formula:

$$\beta = \frac{I_C}{I_B}$$

$$I_B = \frac{I_C}{\beta} = \frac{500\,mA}{100} = 5\,mA$$

QUESTIONS

Calculate the following values:

A. I_C = 2 ampere, β = 20. Find I_B. _____ 2/20

B. I_B = 1 mA, β = 100. Find I_C. ___ .001 × 100

C. I_B = 10 µA, β = 250. Find I_C. ___ .00001 × 250

D. I_B = 0.1 mA, I_C = 7.5 mA. Find β. ___ .0075/.0001

ANSWERS

A. 0.1 ampere, or 100 mA

B. 100 mA

C. 2500 µA, or 2.5 mA

D. 75

19 This problem serves as a summary of the first part of this chapter. You should be able to answer all these questions. Use a separate sheet of paper for your drawing and calculations.

QUESTIONS

A. Draw a transistor circuit utilizing an NPN transistor, a base resistor, a collector resistor, and one battery to supply both base and collector currents. Show the paths of I_B and I_C.

B. Which current controls the other? _____

_____

   C. Which is the larger current, I_B or I_C? _I_c_____

_____

   D. $I_B = 6 \ \mu A$, $\beta = 250$. Find I_C. _.000006 × 250_____

_____

   E. $I_C = 300 \ mA$, $\beta = 50$. Find I_B. _.300/50_____

_____

ANSWERS

A. See Figure 3.17 and Figure 3.18.

B. I_B (base current) controls I_C (collector current).

C. I_C

D. 1.5 mA

E. 6 mA

PROJECT 3.1: The Transistor

OBJECTIVE

The objective of this project is to find β of a particular transistor by setting several values of base current and measuring the corresponding values of collector current. Next, you divide the values of collector current by the values of the base current to determine β. The value of β will be almost the same for all the measured values of current. This demonstrates the operation of a transistor in its linear region of operation, wherein β is almost constant.

GENERAL INSTRUCTIONS

While the circuit is set up, measure the collector voltage for each current value. This demonstrates (experimentally) some points that are covered in future problems. As you perform the project, observe how the collector voltage V_C drops as the collector current increases.

Parts List

❑ One 9 V battery (or a lab power supply)

❑ One multimeter set to μA

❏ One multimeter set to mA

❏ One multimeter set to measure DC voltage

❏ One 10 kΩ resistor

❏ One 510 ohm resistor

❏ One 2N3904 transistor

❏ One breadboard

❏ One 1 MΩ potentiometer

STEP-BY-STEP INSTRUCTIONS

Set up the circuit shown in Figure 3.19 on a breadboard. If you have some experience in building circuits, this schematic (along with the previous parts list) should provide all the information you need to build the circuit. If you need a bit more help building the circuit, look at the photos of the completed circuit in the "Expected Results" section.

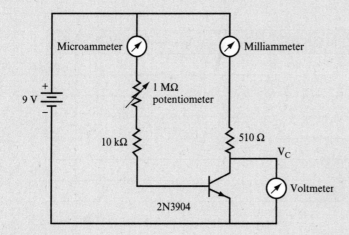

FIGURE 3.19

Follow these steps, recording your measurements in the blank table following the steps.

1. Set the potentiometer to its highest value; this sets I_B to its lowest possible value.

2. Measure and record I_B.

3. Measure and record I_C.

4. Measure and record V_C. This voltage is sometimes referred to as the *collector-emitter voltage* (V_{CE}), because it is taken across the collector-emitter leads if the emitter is connected to ground or the negative of the power supply.

5. Adjust the potentiometer to give the next targeted value of I_B. You do not need to hit these values exactly. For example, for a target of 20 µA, a measured value of 20.4 µA is fine.

6. Measure and record the new values for I_B, I_C, and V_C.

7. Adjust the potentiometer to give the next targeted value of I_B.

8. Measure and record the new values for I_B, I_C, and V_C again.

9. Repeat steps 7 and 8 for each of the targeted values of I_B.

10. For each value of I_B and its corresponding value of I_C, calculate the value of β ($β = I_C/I_B$). The values will vary slightly but will be close to an average. Did you get a consistent β?

Target I_B (µA)	Measured I_B (µA)	I_C (mA)	V_C (volts)	β
Lowest possible				
10				
15				
20				
25				
30				
35				
40				
45				
50				

Save this circuit. You use it later in this chapter in Project 3.2.

EXPECTED RESULTS

Figure 3.20 shows the breadboarded circuit for this project.

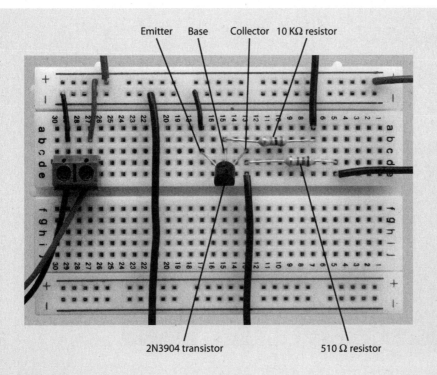

Emitter Base Collector 10 KΩ resistor

2N3904 transistor   510 Ω resistor

FIGURE 3.20

Figure 3.21 shows the test setup for this project.

Compare your measurements with the ones shown in the following table.

Target I_B (µA)	I_B (µA)	I_C (mA)	V_C (volts)	β
Lowest possible	8.7	1.5	8.41	172.4
10	10	1.73	8.3	173
15	15	2.6	7.85	173.3
20	20	3.46	7.43	173
25	25	4.32	6.97	172.8
30	30	5.18	6.54	172.7
35	35	6.06	6.08	173.1
40	40	6.9	5.6	173
45	45	7.76	5.2	172.4
50	50	8.6	4.76	172

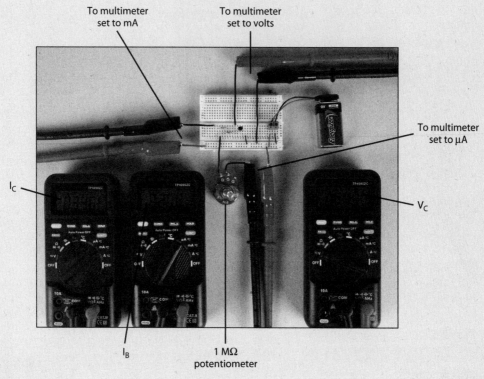

FIGURE 3.21

Don't worry if your results give a different value of β. The manufacturing process that produces transistors can allow variation of the base thickness and doping levels, which causes variation of β in the finished transistors.

20 In Project 3.1, you measured the voltage level at the collector (V_C) and recorded your measurements. Now, examine how to determine the voltage at the collector, when it's not possible to measure the voltage level.

Use the values shown in the circuit in Figure 3.22 to complete the following steps:

1. Determine I_C.

2. Determine the voltage drop across the collector resistor R_C. Call this V_R.

3. Subtract V_R from the supply voltage to calculate the collector voltage.

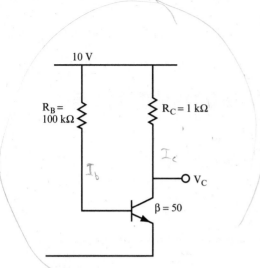

FIGURE 3.22

Here is the first step:

1. To find I_C, you must first find I_B.

$$I_B = \frac{10\,\text{volts}}{100\,\text{k}\Omega} = 0.1\,\text{mA}$$

$$I_C = \beta \times I_B = 50 \times 0.1\,\text{mA} = 5\,\text{mA}$$

Now, perform the next two steps.

QUESTIONS

A. $V_R =$ _____

B. $V_C =$ _____

ANSWERS

A. To find V_R:

$$V_R = R_C \times I_C = 1\,\text{k}\Omega \times 5\,\text{mA} = 5\,\text{volts}$$

B. To find V_C:

$$V_C = V_S - V_R = 10\,\text{volts} - 5\,\text{volts} = 5\,\text{volts}$$

21 Determine parameters for the circuit shown in Figure 3.22 using the value of $\beta = 75$.

QUESTIONS

Calculate the following:

A. $I_C = $ _____

B. $V_R = $ _____

C. $V_C = $ _____

ANSWERS

A. $I_B = \dfrac{10\,\text{volts}}{100\,\text{k}\Omega} = 0.1\,\text{mA}$

   $I_C = 75 \times 0.1\,\text{mA} = 7.5\,\text{mA}$

B. $V_R = 1\,\text{k}\Omega \times 7.5\,\text{mA} = 7.5\,\text{volts}$

C. $V_C = 10\,\text{volts} - 7.5\,\text{volts} = 2.5\,\text{volts}$

22 Determine parameters for the same circuit, using the values of $R_B = 250\,\text{k}\Omega$ and $\beta = 75$.

QUESTIONS

Calculate the following:

A. $I_C = $ _____

B. $V_R = $ _____

C. $V_C = $ _____

ANSWERS

A. $I_B = \dfrac{10\,\text{volts}}{250\,\text{k}\Omega} = \dfrac{1}{25}\,\text{mA}$

   $I_C = 75 \times \dfrac{1}{25}\,\text{mA} = 3\,\text{mA}$

B. $V_R = 1\,k\Omega \times 3\,mA = 3$ volts

C. $V_C = 10$ volts–3 volts = 7 volts

23 From the preceding problems, you can see that you can set V_C to any value by choosing a transistor with an appropriate value of β or by choosing the correct value of R_B.

Now, consider the example shown in Figure 3.23. The objective is to find V_C. Use the steps outlined in problem 20.

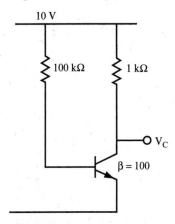

FIGURE 3.23

QUESTIONS

Calculate the following:

A. $I_B = $ _____

$I_C = $ _____

B. $V_R = $ _____

C. $V_C = $ _____

ANSWERS

Your results should be as follows:

A. $I_B = \dfrac{10\text{ volts}}{100\,k\Omega} = 0.1\,mA$

$I_C = 100 \times 0.1\,mA = 10\,mA$

(Continued)

(continued)

> ## ANSWERS
>
> **B.** $V_R = 1\,k\Omega \times 10\,mA = 10$ volts
>
> **C.** $V_C = 10$ volts -10 volts $= 0$ volts.
>
> Here the base current is sufficient to produce a collector voltage of 0 volts and the maximum collector current possible, given the stated values of the collector resistor and supply voltage. This condition is called *saturation*.

24 Look at the two circuits shown in Figure 3.24 and compare their voltages at the point labeled V_C.

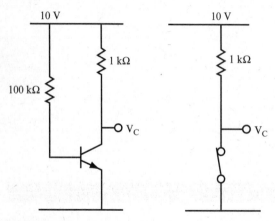

FIGURE 3.24

Consider a transistor that has sufficient base current and collector current to set its collector voltage to 0 volts. Obviously, this can be compared to a closed mechanical switch. Just as the switch is said to be ON, the transistor is also said to be "turned on" (or just ON).

QUESTIONS

A. What can you compare a turned on transistor to? _____

_____

B. What is the collector voltage of an ON transistor?_____

_____

ANSWERS

A. A closed mechanical switch

B. 0 volts

PROJECT 3.2: The Saturated Transistor

OBJECTIVE

Normally, for a transistor, $I_C = \beta \times I_B$. However, this relationship does not hold when a transistor is saturated. The objective of this project is to determine the relationship of the collector current to the base current when a transistor is saturated.

GENERAL INSTRUCTIONS

Using the same breadboarded circuit you built in Project 3.1, set the base current to several values, starting at 90 µA and increasing the base current. Record measurements of the collector current and collector voltage at each value of the base current.

STEP-BY-STEP INSTRUCTIONS

Follow these steps and record your measurements in the blank table following the steps.

1. Set up the circuit you built in Project 3.1.
2. Adjust the potentiometer to set I_B at approximately 90 µA.
3. Measure and record I_B.
4. Measure and record I_C.
5. Measure and record V_C.
6. Adjust the potentiometer slightly to lower its resistance, which sets a larger value of I_B.
7. Measure and record the new values for I_B, I_C, and V_C.
8. Repeat steps 6 and 7 until you reach the lower limit of the potentiometer, which is also the highest value of I_B.

I_B (µA)	I_C (mA)	V_C (volts)

EXPECTED RESULTS

Figure 3.25 shows the test setup for this project with the potentiometer set at its lower limit, providing the highest value of I_B. The test setup is the same as that used in Project 3.1; however, the value of I_B is considerably higher than in Project 3.1.

Compare your measurements to the ones shown in the following table.

I_B (µA)	I_C (mA)	V_C (volts)
91	14.5	1.53
101	15.9	0.843
126	16.9	0.329
150	17.0	0.256
203	17.1	0.211
264	17.1	0.189
389	17.2	0.163
503	17.2	0.149
614	17.2	0.139
780	17.2	0.127
806	17.2	0.126

Your data will probably have slightly different values than shown here but should indicate that I_C stays constant for values of V_C of 0.2 and below, whereas I_B continues to rise. In this region of values the transistor is fully ON (saturated) and I_C can't increase further. This demonstrates that the current gain is not constant for a saturated transistor.

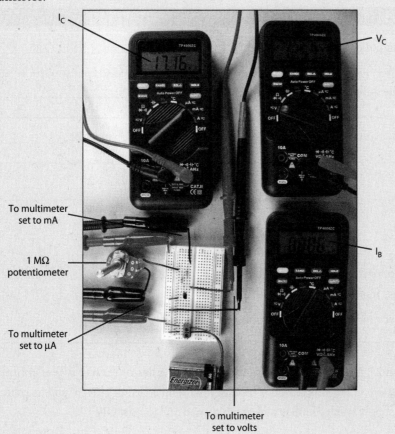

FIGURE 3.25

25 Now, compare the circuits shown in Figure 3.26.

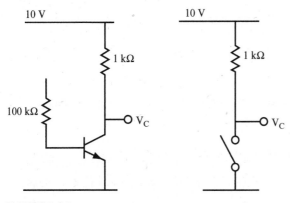

FIGURE 3.26

Because the base circuit is broken (that is, it is not complete), there is no base current flowing.

QUESTIONS

A. How much collector current is flowing? _____

_____

B. What is the collector voltage? _____

_____

C. What is the voltage at the point V_C in the mechanical switch circuit? _____

_____

ANSWERS

A. None.

B. Because there is no current flowing through the 1 kΩ resistor, there is no voltage drop across it, so the collector is at 10 volts.

C. 10 volts because there is no current flowing through the 1 kΩ resistor.

26 From problem 25, it is obvious that a transistor with no collector current is similar to an open mechanical switch. For this reason, a transistor with no collector current and its collector voltage at the supply voltage level is said to be "turned off" (or just OFF).

QUESTION

What are the two main characteristics of an OFF transistor?_____

_____

ANSWER

It has no collector current, and the collector voltage is equal to the supply voltage.

27 Now, calculate the following parameters for the circuit in Figure 3.27, and compare the results to the examples in problems 25 and 26. Again, the objective here is to find V_C.

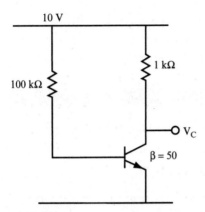

FIGURE 3.27

QUESTIONS

A. $I_B =$ _____

   $I_C =$ _____

B. $V_R =$ _____

C. $V_C =$ _____

ANSWERS

A. $I_B = \dfrac{10\,\text{volts}}{100\,\text{k}\Omega} = 0.1\,\text{mA}$

   $I_C = 50 \times 0.1\text{mA} = 5\text{mA}$

B. $V_R = 1\,\text{k}\Omega \times 5\,\text{mA} = 5\,\text{volts}$

C. $V_C = 10\,\text{volts} - 5\,\text{volts} = 5\,\text{volts}$

NOTE The output voltage in this problem is half of the supply voltage. This condition is important in AC electronics and is covered in Chapter 8.

THE JUNCTION FIELD EFFECT TRANSISTOR (JFET)

28 Up to now, the only transistor described has been the BJT. Another common transistor type is the JFET. Like the BJT, the JFET is used in many switching and amplification applications. The JFET is preferred when a high input impedance circuit is needed.

The BJT has a relatively low input impedance as compared to the JFET. Like the BJT, the JFET is a three-terminal device. The terminals are called the *source*, *drain*, and *gate*. They are similar in function to the emitter, collector, and base, respectively.

QUESTIONS

A. How many terminals does a JFET have, and what are these terminals called? _____

_____

B. Which terminal has a function similar to the base of a BJT? _____

_____

ANSWERS

A. Three, called the source, drain, and gate.

B. The gate has a control function similar to that of the base of a BJT.

29 The basic design of a JFET consists of one type of semiconductor material with a channel made of the opposite type of semiconductor material running through it. If the channel is N material, it is called an N-channel JFET; if it is P material, it is called a P-channel.

Figure 3.28 shows the basic layout of N and P materials, along with their circuit symbols. Voltage on the gate controls the current flow through the drain and source by controlling the effective width of the channel, allowing more or less current to flow. Thus, the voltage on the gate acts to control the drain current, just as the voltage on the base of a BJT acts to control the collector current.

QUESTIONS

A. Which JFET would use electrons as the primary charge carrier for the drain current? _____

_____

B. What effect does changing the voltage on the gate have on the operation of the JFET? _____

_____

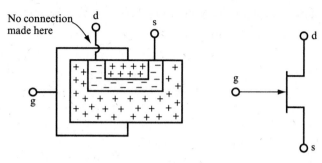

N-channel JFET

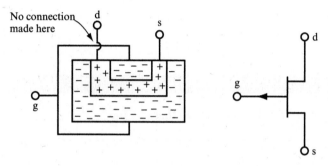

P-channel JFET

FIGURE 3.28

ANSWERS

A. N-channel because N material uses electrons as the majority carrier.

B. It changes the current in the drain. The channel width is controlled electrically by the gate potential.

30 To operate the N-channel JFET, apply a positive voltage to the drain with respect to the source. This allows a current to flow through the channel. If the gate is at 0 volts, the drain current is at its largest value for safe operation, and the JFET is in the ON condition.

When a negative voltage is applied to the gate, the drain current is reduced. As the gate voltage becomes more negative, the current lessens until cutoff, which occurs when the JFET is in the OFF condition.

Figure 3.29 shows a typical biasing circuit for the N-channel JFET. For a P-channel JFET, you must reverse the polarity of the bias supplies.

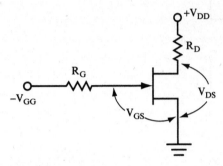

FIGURE 3.29

QUESTION

How does the ON-OFF operation of a JFET compare to that of a BJT?_____

_____

ANSWER

The JFET is ON when there are 0 volts on the gate, whereas you turn the BJT ON by applying a voltage to the base. You turn the JFET OFF by applying a voltage to the gate, and the BJT is OFF when there are 0 volts on the base. The JFET is a "normally ON" device, but the BJT is considered a "normally OFF" device. Therefore, you can use the JFET (like the BJT) as a switching device.

31 When the gate to source voltage is at 0 volts (V_{GS} = 0) for the JFET (refer to Figure 3.29), the drain current is at its maximum (or saturation) value. This means that the N-channel resistance is at its lowest possible value, in the range of 5 to 200 ohms. If R_D is significantly greater than this, the N-channel resistance, r_{DS}, is assumed to be negligible.

QUESTIONS

A. What switch condition would this represent, and what is the drain to source voltage (V_{DS})?_____

_____

B. As the gate becomes more negative with respect to the source, the resistance of the N-channel increases until the cutoff point is reached. At this point, the resistance of the channel is assumed to be infinite. What condition does this represent, and what is the drain to source voltage?_____

_____

C. What does the JFET act like when it is operated between the two extremes of current saturation and current cutoff?_____

_____

ANSWERS

A. Closed switch, $V_{DS} = 0$ volts, or low value

B. Open switch, $V_{DS} = V_{DD}$

C. A variable resistance

INSIDE THE JFET

Now take a closer look at the inside of an N-channel JFET. With 0 volts applied to the gate, the channel is at its widest, and the maximum amount of current can flow between the drain and the source. If you apply a negative voltage to the gate, electrons in the channel are repelled from the negative voltage, forming depletion regions on each side of the channel, which narrows the channel, as shown in the following figure.

Depletion regions

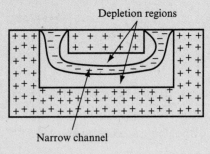

Narrow channel

Continued

continued

Further increasing the negative voltage on the gate repels additional electrons, increasing the width of the depletion region and decreasing the width of the channel. The narrower the channel, the higher its resistance. When you apply high enough negative voltage, the depletion regions completely block the channel, as shown in the following figure, cutting off the flow of current between the drain and source. This voltage is called the *cutoff voltage*.

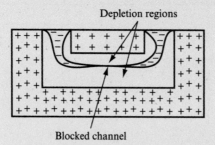

Depletion regions

Blocked channel

Before you connect any JFET to other components in a circuit, you must identify the drain, gate, and source leads (referred to as the JFET's pinout) and determine whether the component is an N-channel or a P-channel JFET.

Transistors are marked with a part number. For example, 2N3819, 2N5951, and 2N5460 are all part numbers of JFETs. However, the part numbers don't tell you much about the JFET. For these three transistors, the 2N3819 and 2N5951 are N channel JFETs, whereas the 2N5460 transistor is a P channel JFET. This is not obvious from the part numbers.

Also, the JFET pinout is not identified on the part number. For example, one N-channel JFET, the 2N3819, uses different leads for the gate, drain, and source than the 2N5951 N-channel JFET, as shown in the following figure.

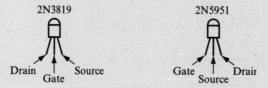

2N3819

Drain   Source
    Gate

2N5951

Gate   Drair
    Source

Refer to the manufacturer's data sheet for the transistor pinout and other characteristics. You can easily look up data sheets on the Internet. Also, you can find links to the data sheets for the JFETs used in this book on the website at `www.buildinggadgets.com/index_datasheets.htm`.

SUMMARY

At this point, it's useful to compare the properties of a mechanical switch with the properties of both types of transistors, as summarized in the following table.

Switch	BJT	JFET
OFF (or open)		
No current.	No collector current.	No drain current.
Full voltage across terminals.	Full supply voltage between collector and emitter.	Full supply voltage between drain and source.
ON (or closed)		
Full current.	Full circuit current.	Full circuit current.
No voltage across terminals.	Collector to emitter voltage is 0 volts.	Drain to source voltage is 0 volts.

The terms ON and OFF are used in digital electronics to describe the two transistor conditions you just encountered. Their similarity to a mechanical switch is useful in many electronic circuits.

In Chapter 4 you learn about the transistor switch in more detail. This is the first step toward an understanding of digital electronics. In Chapter 8 you examine the operation of the transistor when it is biased at a point falling between the two conditions, ON and OFF. In this mode, the transistor can be viewed as a variable resistance and used as an amplifier.

SELF-TEST

The following questions test your understanding of the concepts presented in this chapter. Use a separate sheet of paper for your drawings or calculations. Compare your answers with the answers provided following the test.

1. Draw the symbols for an NPN and a PNP bipolar transistor. Label the terminals of each.

2. Draw the paths taken by the base and collector currents, as shown in Figure 3.30.

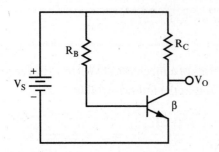

FIGURE 3.30

3. What causes the collector current to flow?_____

_____

4. What is meant by the term *current gain*? What symbol is used for this? What is its algebraic formula? _____

_____

Use the circuit in Figure 3.30 to answer questions 5 through 10.

5. Assume that the transistor is made of silicon. Set $R_B = 27$ kΩ and $V_S = 3$ volts. Find I_B. _____

_____

6. If $R_B = 220$ kΩ and $V_S = 10$ volts. Find I_B. _____

_____

7. Find V_O when $R_B = 100$ kΩ, $V_S = 10$ volts, $R_C = 1$ kΩ, and $\beta = 50$._____

_____

8. Find V_O when $R_B = 200$ kΩ, $V_S = 10$ volts, $R_C = 1$ kΩ, and $\beta = 50$._____

_____

9. Now use these values to find V_O: $R_B = 47$ kΩ, $V_S = 10$ volts, $R_C = 500$ ohms, and $\beta = 65$. _____

_____

10. Use these values to find V_O: $R_B = 68$ kΩ, $V_S = 10$ volts, $R_C = 820$ ohms, and $\beta = 75$._____

_____

11. Draw the symbols for the two types of JFETs and identify the terminals._____

_____

12. What controls the flow of current in both a JFET and a BJT?_____
_____

13. In the JFET common source circuit shown in Figure 3.31, add the correct polarities of the power supplies, and draw the current path taken by the drain current.

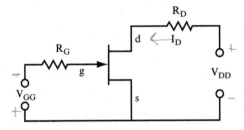

FIGURE 3.31

14. When a base current is required to turn a BJT ON, why is there no gate current for the JFET in the ON state._____
_____

15. Answer the following questions for the circuit shown in Figure 3.32.

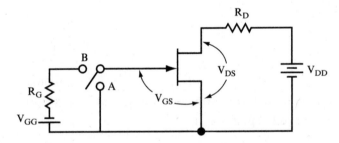

FIGURE 3.32

A. If the switch is at position A, what will the drain current be, and why? _____
_____

B. If the switch is at position B, and the gate supply voltage is of sufficient value to cause cutoff, what will the drain current be, and why? _____
_____

C. What is the voltage from the drain to the source for the two switch positions?
_____

ANSWERS TO SELF-TEST

If your answers do not agree with those that follow, review the problems indicated in parentheses before you go to Chapter 4.

1.	See Figure 3.33. 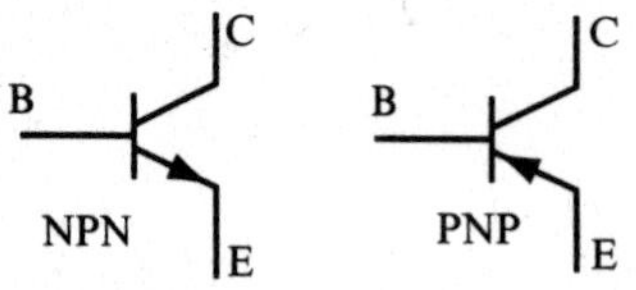 FIGURE 3.33	(problems 4 and 5)
2.	See Figure 3.34. 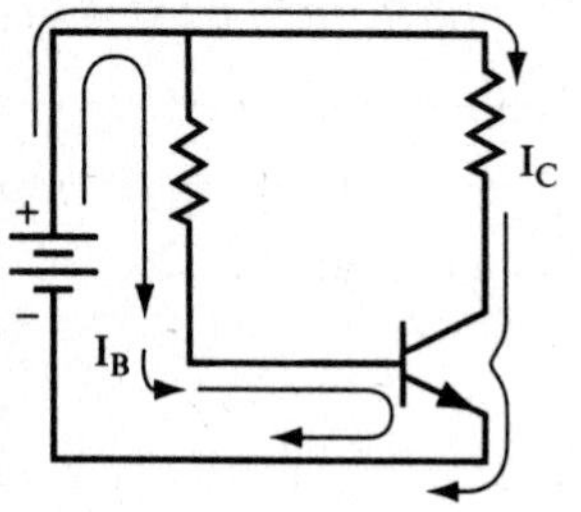 FIGURE 3.34	(problems 13 and 15)
3.	Base current.	(problem 15)
4.	Current gain is the ratio of collector current to base current. It is represented by the symbol β. $\beta = I_C/I_B$.	(problems 16 and 17)
5.	$I_B = \dfrac{(V_S - 0.7\,\text{volt})}{R_B} = \dfrac{(3\,\text{volts} - 0.7\,\text{volt})}{27\,\text{k}\Omega} = \dfrac{2.3\,\text{volts}}{27\,\text{k}\Omega} = 85\mu\text{A}$	(problem 7)
6.	$I_B = \dfrac{10\,\text{volts}}{220\,\text{k}\Omega} = 45.50\,\mu\text{A}$	(problem 7)
7.	5 volts	(problems 20–23)
8.	7.5 volts	(problems 20–23)
9.	3.1 volts	(problems 20–23)
10.	1 volt	(problems 20–23)

11.	See Figure 3.35.	(problem 29)
	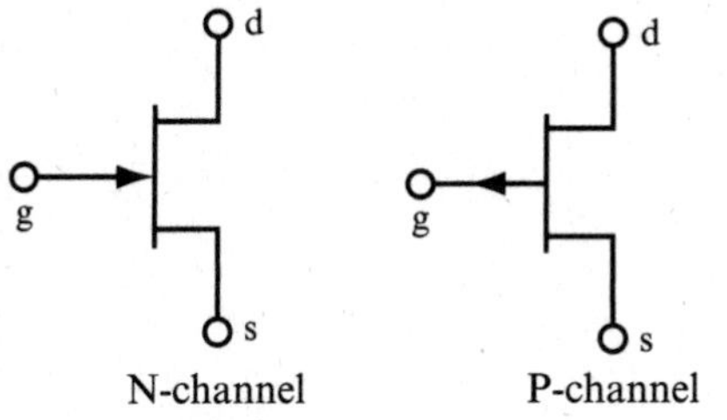 N-channel P-channel **FIGURE 3.35**	
12.	The voltage on the gate controls the flow of drain current, which is similar to the base voltage controlling the collector current in a BJT.	(problem 29)
13.	See Figure 3.36. 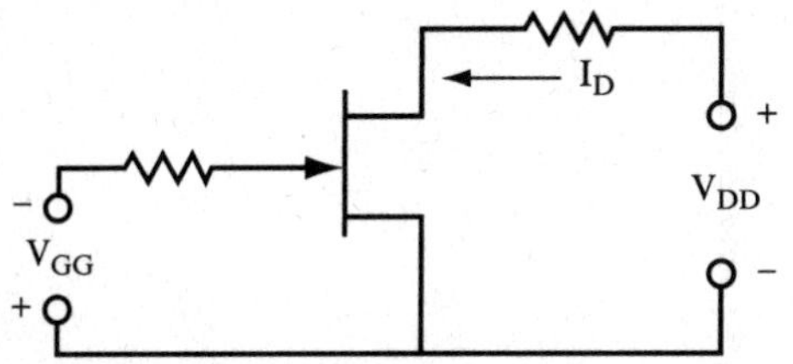 **FIGURE 3.36**	(problem 30)
14.	The JFET is a high-impedance device and does not draw current from the gate circuit. The BJT is a relatively low-impedance device and does, therefore, require some base current to operate.	(problem 28)
15A.	The drain current will be at its maximum value. In this case, it equals V_{DD}/R_D because you can ignore the drop across the JFET. The gate to source voltage is 0 volts, which reduces the channel resistance to a small value close to 0 ohms.	(problem 31)
15B.	The drain current now goes to 0 ampere because the channel resistance is at infinity (very large), which does not allow electrons to flow through the channel.	
15C.	At position A, V_{DS} is approximately 0 volts. At position B, $V_{DS} = V_{DD}$.	

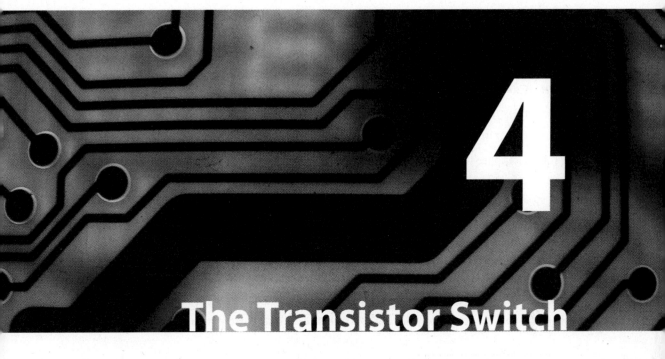

The Transistor Switch

Transistors are everywhere. You can't avoid them as you move through your daily tasks. For example, almost all industrial controls, and even your MP3 player, stereo, and television may use transistors as switches.

In Chapter 3, "Introduction to the Transistor," you saw how a transistor can be turned ON and OFF, similar to a mechanical switch. Computers work with Boolean algebra, which uses only two logic states—TRUE and FALSE. These two states are easily represented electronically by a transistor that is ON or OFF. Therefore, the transistor switch is used extensively in computers. In fact, the logic portions of microprocessors (the brains of computers) consist entirely of transistor switches.

This chapter introduces the transistor's simple and widespread application—switching, with emphasis on the bipolar junction transistor (BJT).

When you complete this chapter, you will be able to do the following:

- Calculate the base resistance, which turns a transistor ON and OFF.

- Explain how one transistor turns another ON and OFF.

- Calculate various currents and resistances in simple transistor switching circuits.

- Calculate various resistances and currents in switching circuits, which contain two transistors.

- Compare the switching action of a junction field effect transistor (JFET) to a BJT.

TURNING THE TRANSISTOR ON

1 Start by examining how to turn a transistor ON by using the simple circuit shown in Figure 4.1. In Chapter 3, R_B was given, and you had to find the value of collector current and voltages. Now, do the reverse. Start with the current through R_C, and find the value of R_B that turns the transistor ON and permits the collector current to flow.

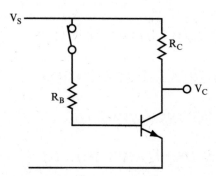

FIGURE 4.1

QUESTION

What current values do you need to know to find R_B? _____

_____

ANSWER

The base and collector currents

2 In this problem circuit, a lamp can be substituted for the collector resistor. In this case, R_C (the resistance of the lamp) is referred to as the *load*, and I_C (the current through the lamp) is called the *load current*.

QUESTIONS

A. Is load current equivalent to base or collector current? _____

_____

B. What is the path taken by the collector current discussed in problem 1? Draw this path on the circuit.

ANSWERS

A. Collector current

B. See Figure 4.2. In this figure, note that the resistor symbol has been replaced by the symbol for an incandescent lamp.

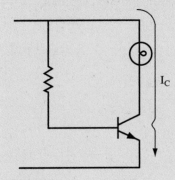

I_C

FIGURE 4.2

3 For the transistor switch to perform effectively as a CLOSED switch, its collector voltage must be at the same voltage as its emitter voltage. In this condition, the transistor is said to be turned ON.

QUESTIONS

A. What is the collector voltage when the transistor is turned ON? _____

_____

B. What other component does an ON transistor resemble? _____

_____

ANSWERS

A. The same as the emitter voltage, which, in this circuit, is 0 volts

B. A closed mechanical switch

NOTE In actual practice, there is a small voltage drop across the transistor from the collector to the emitter. This is actually a saturation voltage and is the smallest voltage drop that can occur across a transistor when it is ON as "hard" as possible. The discussions in this chapter consider this voltage drop to be a negligible value; therefore, the collector voltage is said to be 0 volts. For a quality switching transistor, this is a safe assumption.

4 The circuit in Figure 4.3 shows a lamp with a resistance of 240 ohms in place of R_C.

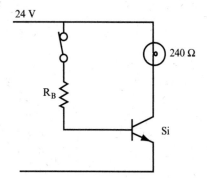

FIGURE 4.3

This figure shows the supply voltage and the collector resistance. Given these two values, using Ohm's law, you can calculate the load current (also called the *collector current*) as follows:

$$I_L = I_C = \frac{V_S}{R_C} = \frac{24 \text{ volts}}{240 \text{ ohms}} = 100 \text{ mA}$$

Thus, 100 mA of collector current must flow through the transistor to fully illuminate the lamp. As you learned in Chapter 3 collector current does not flow unless base current flows.

QUESTIONS

A. Why do you need base current? _____

_____

B. How can you make base current flow? _____

_____

ANSWERS

A. To enable collector current to flow so that the lamp lights up

B. By closing the mechanical switch in the base circuit

5 You can calculate the amount of base current flowing. Assume that $\beta = 100$.

QUESTION

A. What is the value of the base current I_B? _____

_____

ANSWER

$$I_B = \frac{I_C}{\beta} = \frac{100\,\text{mA}}{100} = 1\,\text{mA}$$

6 The base current flows in the direction shown in Figure 4.4. Base current flows through the base-emitter junction of the transistor as it does in a forward-biased diode.

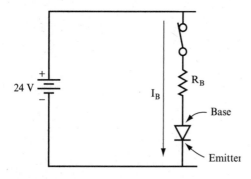

FIGURE 4.4

QUESTIONS

A. What is the voltage drop across the base-emitter diode? _____

_____

B. What is the voltage drop across R_B? _____

_____

ANSWERS

A. 0.7 volt because it is a silicon transistor

B. 24 volts if the 0.7 is ignored; 23.3 volts if it is not

7 The next step is to calculate R_B. The current flowing through R_B is the base current I_B, and you determined the voltage across it in problem 6.

QUESTION

Calculate R_B. _____

_____

ANSWER

$$R_B = \frac{23.3 \text{ volts}}{1 \text{ mA}} = 23,300 \text{ ohms}$$

Figure 4.5 shows the final circuit, including the calculated current and resistance values.

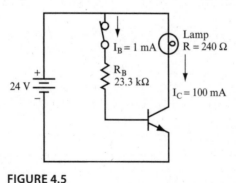

FIGURE 4.5

8 Use the following steps to calculate the values of I_B and R_B needed to turn a transistor ON:

1. Determine the required collector current.

2. Determine the value of β.

3. Calculate the required value of I_B from the results of steps 1 and 2.

4. Calculate the required value of R_B.

5. Draw the final circuit.

Now, assume that $V_S = 28$ volts, that you are using a lamp requiring 50 mA of current, and that $\beta = 75$.

QUESTIONS

A. Calculate I_B. _____

_____

B. Determine R_B. _____

_____

ANSWERS

A. The collector current and β were given. Thus:

$$I_B = \frac{I_C}{\beta} = \frac{50\,\text{mA}}{75} = 0.667\,\text{mA}$$

B. $R_B = \dfrac{28\ \text{volts}}{0.667\,\text{mA}} = 42\text{k}\Omega$

This calculation ignores V_{BE}.

9 Now, assume that $V_S = 9$ volts, that you are using a lamp requiring 20 mA of current, and that $\beta = 75$.

QUESTION

Calculate R_B. _____

_____

ANSWER

$R_B = 31.1k\Omega$

In this calculation, V_{BE} is included.

10 In practice, if the supply voltage is much larger than the 0.7-volt drop across the base-emitter junction, you can simplify your calculations by ignoring the 0.7-volt drop, and assume that all the supply voltage appears across the base resistor R_B. (Most resistors are only accurate to within +/- 5 percent of their stated value anyway.) If the supply voltage is less than 10 volts, however, you shouldn't ignore the 0.7-volt drop across the base-emitter junction.

QUESTIONS

Calculate R_B for the following problems, ignoring the voltage drop across the base-emitter junction, if appropriate.

A. A 10-volt lamp that draws 10 mA. $\beta = 100$. _____

_____

B. A 5-volt lamp that draws 100 mA. $\beta = 50$. _____

_____

ANSWERS

A. $I_B = \dfrac{10\,mA}{100} = 0.1\,mA$

$R_B = \dfrac{10\,volts}{0.1\,mA} = 100k\Omega$

B. $I_B = \dfrac{100\,mA}{50} = 2\,mA$

$R_B = \dfrac{(5\,volts - 0.7\,volts)}{2\,mA} = \dfrac{4.3\,volts}{2\,mA} = 2.15k\Omega$

TURNING OFF THE TRANSISTOR

11 Up to now, you have concentrated on turning the transistor ON, thus making it act like a closed mechanical switch. Now you focus on turning it OFF, thus making it act like an open mechanical switch. If the transistor is OFF, no current flows through the load (that is, no collector current flows).

QUESTIONS

A. When a switch is open, are the two terminals at different voltages or at the same voltage? _____

_____

B. When a switch is open, does current flow? _____

_____

C. For a transistor to turn OFF and act like an open switch, how much base current is needed? _____

_____

ANSWERS

A. At different voltages, the supply voltage and ground voltage.

B. No.

C. The transistor is OFF when there is no base current.

12 You can be sure that there is no base current in the circuit shown in Figure 4.6 by opening the mechanical switch.

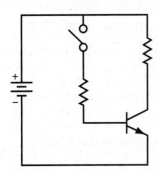

FIGURE 4.6

To ensure that the transistor remains OFF when the base is not connected to the supply voltage, you add a resistor (labeled R_2 in Figure 4.7) to the circuit. The base of the transistor connects to ground or 0 volts through this resistor. Therefore, no base current can possibly flow.

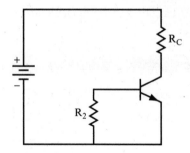

FIGURE 4.7

QUESTIONS

A. Why doesn't current flow from the supply voltage to the base-emitter junction? ___

_____

B. How much current flows from collector to base? _____

_____

C. Why doesn't current flow from collector to base through R_2 ground?_____

_____

D. Why is the transistor base at 0 volts when R_2 is installed?_____

_____

ANSWERS

A. There is no current path from the supply voltage through the base-emitter junction. Thus, there is no base current flowing.

B. None at all.

C. The internal construction of the transistor prevents this, because the collector-to-base junction is basically a reverse-biased diode.

D. Because there is no current through R_2, there is no voltage drop across R_2 and, therefore, the transistor base is at ground (0 volts).

13 Because no current is flowing through R_2, you can use a wide range of resistance values. In practice, the values you find for R_2 are between 1 kΩ and 1 MΩ.

QUESTION

Which of the following resistor values would you use to keep a transistor turned off?
1 ohm, 2 kΩ, 10 kΩ, 20 kΩ, 50 kΩ, 100 kΩ, 250 kΩ, and 500 kΩ. _____

_____

ANSWER

They would all be suitable except the 1 ohm because the rest are all above 1 kΩ and
below 1 MΩ.

14 Figure 4.8 shows a circuit using both R_1 and R_2. Note that the circuit includes a two-position switch that you can use to turn the transistor ON or OFF.

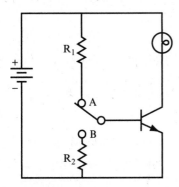

FIGURE 4.8

QUESTIONS

A. As shown in Figure 4.8, is the transistor ON or OFF? _____

_____

B. Which position, A or B, can cause the collector current to be 0 amperes? _____

_____

ANSWERS

A. ON—the base-to-emitter diode is forward-biased. Therefore, base current can flow.

B. Position B—the base is tied to ground. Therefore, no base current can flow, and
the transistor is OFF.

WHY TRANSISTORS ARE USED AS SWITCHES

15 You can use the transistor as a switch (as you saw in the previous problems) to perform simple operations such as turning a lamp current on and off. Although often used between a mechanical switch and a lamp, there are other uses for the transistor.

Following are a few other examples that demonstrate the advantages of using a transistor in a circuit as a switch:

- *Example 1*—Suppose you must put a lamp in a dangerous environment, such as a radioactive chamber. Obviously, the switch to operate the lamp must be placed somewhere safe. You can simply use a switch outside the chamber to turn the transistor switch ON or OFF.

- *Example 2*—If a switch controls equipment that requires large amounts of current, then that current must flow through the wires that run between the switch and the lamp. Because the transistor switch can be turned ON or OFF using low voltages and currents, you can connect a mechanical switch to the transistor switch using small, low-voltage wire and, thereby, control the larger current flow. If the mechanical switch is any distance from the equipment you're controlling, using low-voltage wire can save you time and money.

- *Example 3*—A major problem with switching high current in wires is that the current induces interference in adjacent wires. This can be disastrous in communications equipment such as radio transceivers. To avoid this, you can use a transistor to control the larger current from a remote location, reducing the current needed at the switch located in the radio transceiver.

- *Example 4*—In mobile devices (such as a radio-controlled airplane), using transistor switches minimizes the power, weight, and bulk required.

- *Example 5*—When you use a sensor to activate devices, the sensor provides a low current to the transistor, which then acts as a switch controlling the larger current needed to power the equipment. An everyday example is a sensor that detects a light beam across a doorway. When the beam is blocked by a person or object passing through, the sensor stops generating a current, switching a transistor OFF, which activates a buzzer.

QUESTION

What features mentioned in these examples make using transistors as switches desirable? _____

_____

ANSWER

The switching action of a transistor can be directly controlled by an electrical signal, as well as by a mechanical switch in the base circuit. This provides a lot of flexibility for the design and allows for simple electrical control. Other factors include safety, reduction of interference, remote switching control, and lower design costs.

16 The following examples of transistor switching demonstrate some other reasons for using transistors:

- *Example 1*—You can control the ON and OFF times of a transistor accurately, whereas mechanical devices are not accurate. This is important in applications such as photography, where it is necessary to expose a film or illuminate an object for a precise period of time. In these types of uses, transistors are much more accurate and controllable than any other device.

- *Example 2*—A transistor can be switched ON and OFF millions of times a second and will last for many years. In fact, transistors are one of the longest lasting and most reliable components known, whereas mechanical switches usually fail after a few thousand operations.

- *Example 3*—The signals generated by most industrial control devices are digital. These control signals can be simply a high or low voltage, which is ideally suited to turning transistor switches ON or OFF.

- *Example 4*—Modern manufacturing techniques enable the miniaturization of transistors to such a great extent that many of them (even hundreds of millions) can be fabricated into a single silicon chip. Silicon chips on which transistors (and other electronic components) have been fabricated are called *integrated circuits* (ICs). ICs are little, flat, black plastic components built into almost every mass-produced electronic device and are the reason that electronic devices continue to get smaller and lighter.

QUESTION

What other features, besides the ones mentioned in the previous problem, are demonstrated in the examples given here?_____

_____

ANSWER

Transistors can be accurately controlled, have high-speed operation, are reliable, have a long life, are small, have low power consumption, can be manufactured in large numbers at low cost, and are extremely small.

17 At this point, consider the idea of using one transistor to turn another one ON and OFF, and of using the second transistor to operate a lamp or other load. (This idea is explored in the next section of this chapter.)

If you must switch many high-current loads, then you can use one switch that controls several transistors simultaneously.

QUESTIONS

A. With the extra switches added, is the current that flows through the main switch more or less than the current that flows through the load? _____

_____

B. What effect do you think the extra transistor has on the following?

1. Safety _____

_____

2. Convenience to the operator _____

_____

3. Efficiency and smoothness of operation _____

_____

ANSWERS

A. Less current flows through the main switch than through the load.

B.

1. It increases safety and allows the operator to stay isolated from dangerous situations.

2. Switches can be placed conveniently close together on a panel, or in the best place for an operator, rather than the switch position dictating the operator position.

3. One switch can start many things, such as in a master lighting panel in a television studio or theater.

18 This problem reviews your understanding of the concepts presented in problems 15, 16, and 17.

QUESTION

Indicate which of the following are good reasons for using a transistor as a switch:

A. To switch equipment in a dangerous or inaccessible area on and off

B. To switch low currents or voltages

C. To lessen the electrical noise that might be introduced into communication and other circuits

D. To increase the number of control switches

E. To use a faster, more reliable device than a mechanical switch _____

_____

ANSWER

A, C, and E.

PROJECT 4.1: The Transistor Switch

OBJECTIVE

The objective of this project is to demonstrate how light can switch a transistor ON or OFF to control a device.

GENERAL INSTRUCTIONS

This project uses two breadboarded circuits. The circuit shown on the left side of Figure 4.9 is used to generate infrared light. Another circuit, shown on the right side of Figure 4.9, switches on a buzzer when the infrared light is blocked by an object.

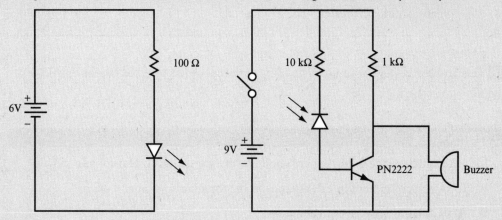

FIGURE 4.9

The infrared light in this project is generated by a light-emitting diode (LED). In an LED, a current runs through a PN junction that generates light. This same process occurs with all diodes. Infrared LEDs are simply diodes with a transparent case that enables the infrared light to show through. LEDs also have a PN junction made with semiconductor material that produces a large amount of infrared light. Figure 4.10 shows a typical LED and its schematic symbol, the symbol for a diode with arrows pointing outward, indicating that light is generated.

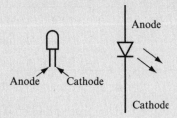

FIGURE 4.10

In this project, a photodiode detects the infrared light. When light strikes a PN junction in a photodiode (or any diode), a current is generated. Infrared photodiodes also have a transparent case and junction material that produces a large current when it absorbs infrared light. Figure 4.11 shows a typical photodiode and its schematic symbol consisting of the symbol for a diode with arrows pointing inward, indicating that light is absorbed.

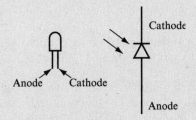

FIGURE 4.11

When the circuits are set up, the buzzer sounds whenever the infrared light is blocked from the photodiode.

Parts List

- ❑ One 9-volt battery.
- ❑ One 6-volt battery pack (4 AA batteries).
- ❑ Two battery snap connectors.
- ❑ One 100-ohm, 0.5-watt resistor.
- ❑ One 1 kΩ, 0.25-watt resistor.
- ❑ One 10 kΩ, 0.25-watt resistor.
- ❑ Two breadboards.
- ❑ Two terminal blocks.
- ❑ One piezoelectric buzzer with a minimum operating voltage of 3 volts DC. Using a buzzer with pins (such as part # SE9-2202AS by Shogyo International) enables you to insert the buzzer directly into the breadboard. If you use a buzzer with wire leads (such as part # PK-27N26WQ by Mallory), you need another terminal block.
- ❑ One infrared LED.
- ❑ One infrared photodiode.
- ❑ One PN2222 transistor. Figure 4.12 shows the pinout diagram for the PN2222.

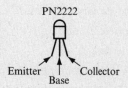

PN2222

Emitter   Collector

Base

FIGURE 4.12

STEP-BY-STEP INSTRUCTIONS

Set up the circuits shown in Figure 4.9. If you have some experience in building circuits, this schematic (along with the previous parts list) should provide all the information you need to build the circuit. If you need a bit more help building the circuit, look at the photos of the completed circuit in the "Expected Results" section.

Carefully check your circuit against the diagram, especially the connection of the long and short leads to the LED and photodiode. The LED is connected so that it is forward-biased, whereas the photodiode is connected so that it is reverse-biased, as indicated by the direction of the schematic symbols in the circuit diagrams.

1. Align the rounded top of the LED toward the rounded top of the photodiode with the circuit boards a few feet apart from each other. (If you use a typical LED and photodiode, you must bend their leads to align them.) Note that the rounded top of both the LED and photodiode shown in Figures 4.10 and 4.11 contain a lens to emit or absorb light. Some LEDs and photodiodes have lenses on the side, instead of on the top. If it isn't obvious where the lens is in your components, check the manufacturer's data sheet.

2. Turn on the power switch. When the power switch is on, the buzzer should sound whenever the photodiode does not sense infrared light.

3. Bring the circuits close enough together so that the buzzer shuts off.

4. Block the infrared light; the buzzer should turn on.

EXPECTED RESULTS

Figure 4.13 shows the breadboarded buzzer circuit for this project.

Buzzer   Photodiode   Collector   PN2222 transistor   Emitter

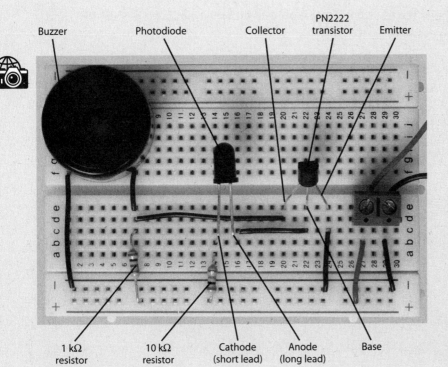

1 kΩ resistor   10 kΩ resistor   Cathode (short lead)   Anode (long lead)   Base

FIGURE 4.13

Figure 4.14 shows the breadboarded LED circuit for this project.

LED

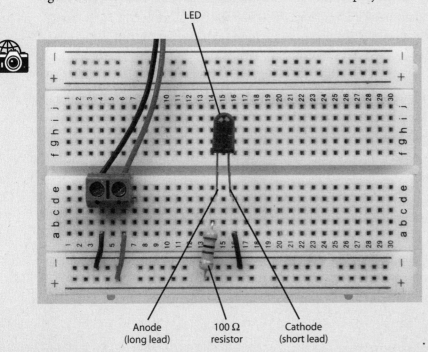

Anode (long lead)   100 Ω resistor   Cathode (short lead)

FIGURE 4.14

Figure 4.15 shows the test setup for this project with the rounded top of the LED and photodiode aligned toward each other.

FIGURE 4.15

The photodiode is connected to the base of a transistor. Therefore, current generated by the photodiode turns the transistor ON. When the transistor is ON, V_C is about 0 volts, turning off the buzzer. When the infrared light is blocked, the photodiode stops generating current, which turns OFF the transistor, increasing V_C, which turns on the buzzer.

These circuits work with the LED and photodiode about 7 inches apart. With more complicated photo detectors that have circuitry to amplify the detected signal, this technique can work over several feet. One common application of this technique is a buzzer that sounds when a shopper enters a store, blocking the light, setting off a sound, and alerting the sales staff.

19 Many types of electronic circuits contain multiple switching transistors. In this type of circuit, one transistor is used to switch others ON and OFF. To illustrate how this works, again consider the lamp as the load and the mechanical switch as the actuating element. Figure 4.16 shows a circuit that uses two transistors to turn a lamp on or off.

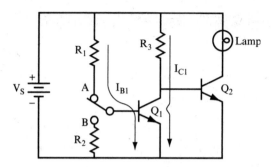

FIGURE 4.16

When the switch is in position A, the base-emitter junction of Q_1 is forward-biased. Therefore, base current (I_{B1}) flows through R_1 and through the base-emitter diode of Q_1, turning the transistor ON. This causes the collector current (I_{C1}) to flow through Q_1 to ground, and the collector voltage drops to 0 volts, just as if Q_1 were a closed switch. Because the base of Q_2 is connected to the collector of Q_1, the voltage on the base of Q_2 also drops to 0 volts. This ensures that Q_2 is turned OFF and the lamp remains unlit.

Now, flip the switch to position B, as shown in Figure 4.17. The base of Q_1 is tied to ground, or 0 volts, turning Q_1 OFF. Therefore, no collector current can flow through Q_1. A positive voltage is applied to the base of Q_2, and the emitter-base junction of Q_2 is forward-biased. This enables current to flow through R_3 and the emitter-base junction of Q_2, which turns Q_2 ON, allowing collector current (I_{C2}) to flow, and the lamp is illuminated.

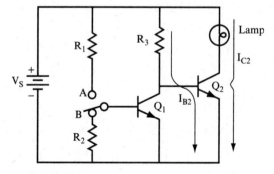

FIGURE 4.17

Now that you have read the descriptions of how the circuit works, answer the following questions. First assume that the switch is in position A, as shown in Figure 4.16.

QUESTIONS

A. What effect does I_{B1} have on transistor Q_1? _____

_____

B. What effect does turning Q_1 ON have on the following?

   1. Collector current I_{C1} _____

_____

   2. Collector voltage V_{C1} _____

_____

C. What effect does the change to V_{C1} covered in the previous question have on the following?

   1. The base voltage of Q_2 _____

_____

   2. Transistor Q_2 (that is, is it ON or OFF) _____

_____

D. Where does the current through R_3 go? _____

_____

E. In this circuit is the lamp on or off? _____

_____

ANSWERS

A. I_{B1}, along with a portion of V_S (0.7 volts if the transistor is silicon), turns Q_1 ON.

B. (1) I_{C1} flows; (2) V_{C1} drops to 0 volts.

C. (1) base of Q_2 drops to 0 volts; (2) Q_2 is OFF.

D. I_{C1} flows through Q_1 to ground.

E. Off.

20 Now, assume that the switch is in the B position, as shown in Figure 4.17, and answer these questions.

QUESTIONS

A. How much base current I_{B1} flows into Q_1? _____

_____

B. Is Q_1 ON or OFF? _____

_____

C. What current flows through R_3? _____

_____

D. Is Q_2 ON or OFF? _____

_____

E. Is the lamp on or off? _____

_____

ANSWERS

A. None

B. OFF

C. I_{B2}

D. ON

E. On

21 Refer to the circuit in Figures 4.16 and 4.17. Now, answer these questions assuming the supply voltage is 10 volts.

QUESTIONS

A. Is the current through R_3 ever divided between Q_1 and Q_2? Explain. _____

_____

_____

B. What is the collector voltage of Q_2 with the switch in each position? _____

_____

C. What is the collector voltage of Q_1 with the switch in each position? _____

_____

ANSWERS

A. No. If Q_1 is ON, all the current flows through it to ground as collector current. If Q_1 is OFF, all the current flows through the base of Q_2 as base current.

B. In position A, 10 volts because it is OFF.

In position B, 0 volts because it is ON.

C. In position A, 0 volts because it is ON.

In position B, the collector voltage of Q_1 equals the voltage drop across the forward-biased base-emitter junction of Q_2, because the base of Q_2 is in parallel with the collector of Q_1. The voltage drop across the forward-biased base-emitter junction does not rise to 10 volts, but can rise only to 0.7 volt if Q_2 is made of silicon.

22 Now, calculate the values of R_1, R_2, and R_3 for this circuit. The process is similar to the one you used before, but you must expand it to deal with the second transistor. This is similar to the steps you used in problem 8. Follow these steps to calculate R_1, R_2, and R_3:

1. Determine the load current I_{C2}.

2. Determine β for Q_2. Call this β_2.

3. Calculate I_{B2} for Q_2. Use $I_{B2} = I_{C2}/\beta_2$.

4. Calculate R_3 to provide this base current. Use $R_3 = V_S/I_{B2}$.

5. R_3 is also the load for Q_1 when Q_1 is ON. Therefore, the collector current for Q_1 (I_{C1}) has the same value as the base current for Q_2, as calculated in step 3.

6. Determine β_1, the β for Q_1.

7. Calculate the base current for Q_1. Use $I_{B1} = I_{C1}/\beta_1$.

8. Find R_1. Use $R_1 = V_S/I_{B1}$.

9. Choose R_2. For convenience, let $R_2 = R_1$.

Continue to work with the same circuit shown in Figure 4.18. Use the following values:

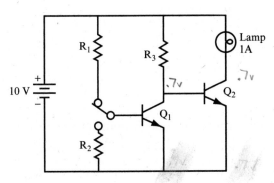

FIGURE 4.18

- A 10-volt lamp that draws 1 ampere; therefore $V_S = 10$ volts, $I_{C2} = 1$ A.
- $\beta_2 = 20$, $\beta_1 = 100$

Ignore any voltage drops across the transistors.

QUESTIONS

Calculate the following:

A. Find I_{B2} as in step 3.

$I_{B2} =$ _____

B. Find R_3 as in step 4.

$R_3 =$ _____

C. Calculate the load current for Q_1 when it is ON, as shown in Step 5.

$I_{C1} =$ _____

D. Find the base current for Q_1.

$I_{B1} =$ _____

E. Find R_1 as in step 8.

$R_1 =$ _____

F. Choose a suitable value for R_2.

$R_2 =$ _____

ANSWERS

The following answers correspond to the steps.

A.

1. I_{C2} is given as 1 ampere.

2. $\beta_2 = 20$ (given). This is a typical value for a transistor that would handle 1 ampere.

3. $I_{B2} = \dfrac{1\,\text{ampere}}{20} = 50\,\text{mA}$

B.

4. $R_3 = \dfrac{10\,\text{volts}}{50\,\text{mA}} = 200\Omega$

    Note that the 0.7 volt base-emitter drop has been ignored.

C.

5. $I_{C1} = I_{B2} = 50\,\text{mA}$

D.

6. $\beta_1 = 100$

7. $I_{B1} = \dfrac{50\,\text{mA}}{100} = 0.5\,\text{mA}$

E.

8. $R_1 = \dfrac{10\,\text{volts}}{0.5\,\text{mA}} = 20\text{k}\Omega$

    Again, the 0.7-volt drop is ignored.

F.

9. For convenience, choose a value for R_2 that is the same as R_1, or 20 kΩ. This reduces the number of different components in the circuit. The fewer different components you have in a circuit, the less components you must keep in your parts bin. You could, of course, choose any value between 1 kΩ and 1 MΩ.

23 Following the same procedure, and using the same circuit shown in Figure 4.18, work through this example. Assume that you are using a 28-volt lamp that draws 560 mA, and that $\beta_2 = 10$ and $\beta_1 = 100$.

QUESTIONS

Calculate the following:

A. $I_{B2} =$ _____

B. $R_3 =$ _____

C. $I_{C1} =$ _____

D. $I_{B1} =$ _____

E. $R_1 =$ _____

F. $R_2 =$ _____

ANSWERS

A. 56 mA

B. 500 ohms

C. 56 mA

D. 0.56 mA

E. 50 kΩ

F. 50 kΩ by choice

THE THREE-TRANSISTOR SWITCH

24 The circuit shown in Figure 4.19 uses three transistors to switch a load on and off. In this circuit, Q_1 is used to turn Q_2 ON and OFF, and Q_2 is used to turn Q_3 ON and OFF. The calculations are similar to those you performed in the last few problems, but a few additional steps are required to deal with the third transistor. Use the circuit diagram in Figure 4.19 to determine the answers to the following questions.

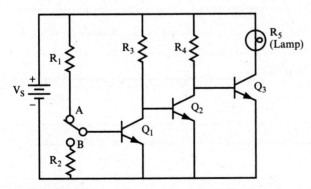

FIGURE 4.19

QUESTIONS

Assume that the switch is in position A.

A. Is Q_1 ON or OFF? _____

_____

B. Is Q_2 ON or OFF?_____

_____

C. Where is current through R_4 flowing?_____

_____

D. Is Q_3 ON or OFF? _____

_____

ANSWERS

A. ON

B. OFF

C. Into the base of Q_3

D. ON

25 Now use the same circuit as in problem 24.

QUESTIONS

Assume that the switch is in position B.

A. Is Q_1 ON or OFF? _____

_____

B. Is Q_2 ON or OFF? _____

_____

C. Where is the current through R_4 flowing?_____

_____

D. Is Q_3 ON or OFF?_____

_____

E. Which switch position turns on the lamp?_____

_____

F. How do the ON/OFF positions for the switch in the three-transistor switch differ from the ON/OFF positions for the switch in the two-transistor switch circuit?____

_____

ANSWERS

A. OFF.

B. ON.

C. Through Q_2 to ground.

D. OFF.

E. Position A.

F. The positions are opposite. Therefore, if a circuit controls lamps with two transistors and another circuit controls lamps with three transistors, flipping the switch that controls both circuits would change which lamps (or which other loads) are on.

26 Work through this example using the same equations you used for the two-transistor switch in problem 22. The steps are similar but with a few added steps, as shown here:

1. Find the load current. This is often given.

2. Determine the current gain of Q_3. This is β_3 and usually it is a given value.

3. Calculate I_{B3}. Use $I_{B3} = I_{C3}/\beta_3$.

4. Calculate R_4. Use $R_4 = V_S/I_{B3}$.

5. Assume $I_{C2} = I_{B3}$.

6. Find β_2. Again this is a given value.

7. Calculate I_{B2}. Use $I_{B2} = I_{C2}/\beta_2$.

8. Calculate R_3. Use $R_3 = V_S/I_{B2}$.

9. Assume $I_{C1} = I_{B2}$.

10. Find β_1.

11. Calculate I_{B1}. Use $I_{B1} = I_{C1}/\beta_1$.

12. Calculate R_1. Use $R_1 = V_s/I_{B1}$.

13. Choose R_2.

For this example, use a 10-volt lamp that draws 10 amperes. Assume that the βs of the transistors are given in the manufacturer's data sheets as $\beta_1 = 100$, $\beta_2 = 50$, and $\beta_3 = 20$. Now, work through the steps, checking the answers for each step as you complete it.

QUESTIONS

Calculate the following:

A. I_{B3} = _____

B. R_4 = _____

C. I_{B2} = _____

D. R_3 = _____

E. I_{B1} = _____

F. R_1 = _____

G. R_2 = _____

ANSWERS

The answers here correspond to the steps.

A.

1. The load current is given as 10 amperes.

2. β_3 is given as 20.

3. $I_{B3} = \dfrac{I_{C3}}{\beta_3} = \dfrac{10 \text{ amperes}}{20} = 0.5 \text{ ampere} = 500\,\text{mA}$

B.

4. $R_4 = \dfrac{10 \text{ volts}}{500\,\text{mA}} = 20\,\text{ohms}$

C.

5. $I_{C2} = I_{B3} = 500\,\text{mA}$

6. β_2 is given as 50.

7. $I_{B2} = \dfrac{I_{C2}}{\beta_2} = \dfrac{500\,\text{mA}}{50} = 10\,\text{mA}$

D.

8. $R_3 = \dfrac{10\text{ volts}}{10\,\text{mA}} = 1\text{k}\Omega$

E.

9. $I_{C1} = I_{B2} = 10\,\text{mA}$

10. β_1 is given as 100.

11. $I_{B1} = \dfrac{I_{C1}}{\beta_1} = \dfrac{10\,\text{mA}}{100} = 0.1\,\text{mA}$

F.

12. $R_1 = \dfrac{10\text{ volts}}{0.1\,\text{mA}} = 100\text{k}\Omega$

G.

13. R_2 can be chosen to be 100 kΩ also.

---

27 Determine the values in the same circuit for a 75-volt lamp that draws 6 amperes. Assume that $\beta_3 = 30$, $\beta_2 = 100$, and $\beta_1 = 120$.

QUESTIONS

Calculate the following values using the steps in problem 26:

A. $I_{B3} = $ _____

B. $R_4 = $ _____

C. $I_{B2} = $ _____

D. $R_3 = $ _____

E. $I_{B1} = $ _____

F. $R_1 = $ _____

G. $R_2 = $ _____

ANSWERS

A. 200 mA

B. 375 Ω

C. 2 mA

D. 37.5 kΩ

E. 16.7 μA

F. 4.5 MΩ

G. Choose $R_2 = 1$ MΩ

ALTERNATIVE BASE SWITCHING

28 In the examples of transistor switching, the actual switching was performed using a small mechanical switch placed in the base circuit of the first transistor. This switch has three terminals and switches from position A to position B. (This is a single-pole, double-throw switch.) This switch does not have a definite ON or OFF position, as does a simple ON-OFF switch.

QUESTION

Why couldn't a simple ON-OFF switch with only two terminals have been used with these examples? _____

_____

ANSWER

An ON-OFF switch is either open or closed, and cannot switch between position A and position B, as shown earlier in Figure 4.19.

29 If you connect R_1, R_2, and a switch together, as shown in Figure 4.20, you can use a simple ON-OFF switch with only two terminals. (This is a single-pole, single-throw switch.)

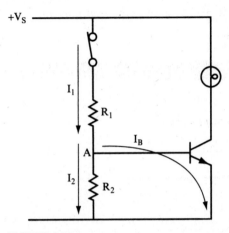

FIGURE 4.20

QUESTIONS

A. When the switch is open, is Q_1 ON or OFF? _____

_____

B. When the switch is closed, is the lamp ON or Off? _____

_____

ANSWERS

A. OFF

B. ON

30 When the switch is closed, current flows through R_1. However, at point A in Figure 4.20, the current divides into two paths. One path is the base current I_B, and the other is marked I_2.

QUESTION

How could you calculate the total current I_1?_____

_____

ANSWER

$I_1 = I_B + I_2$

31 The problem now is to choose the values of both R_1 and R_2 so that when the current divides, there is sufficient base current to turn Q_1 ON.

QUESTION

Consider this simple example. Assume the load is a 10-volt lamp that needs 100 mA of current and $\beta = 100$. Calculate the base current required.

$I_B = $_____

ANSWER

$$I_B = \frac{100\,mA}{100} = 1\,mA$$

32 After the current I_1 flows through R_1, it must divide, and 1 mA of it becomes I_B. The remainder of the current is I_2. The difficulty at this point is that there is no unique value for either I_1 or I_2. In other words, you could assign them almost any value. The only restriction is that both must permit 1 mA of current to flow into the base of Q_1.

You must make an arbitrary choice for these two values. Based on practical experience, it is common to set I_2 to be 10 times greater than I_B. This split makes the circuits work reliably and keeps the calculations easy:

$I_2 = 10\,I_B$

$I_1 = 11\,I_B$

QUESTION

In problem 31 you determined that $I_B = 1$ mA. What is the value of I_2? _____

_____

ANSWER

$I_2 = 10$ mA

33 Now you can calculate the value of R_2. The voltage across R_2 is the same as the voltage drop across the base-emitter junction of Q_1. Assume that the circuit uses a silicon transistor, so this voltage is 0.7 volt.

QUESTIONS

A. What is the value of R_2? _____

_____

B. What is the value R_1? _____

_____

ANSWERS

A. $R_2 = \dfrac{0.7 \text{ volt}}{10 \text{ mA}} = 70 \text{ ohms}$

B. $R_1 = \dfrac{(10 \text{ volts} - 0.7 \text{ volt})}{11\text{mA}} = \dfrac{9.3 \text{ volts}}{11\text{mA}} = 800 \text{ ohms (approximately)}$

You can ignore the 0.7 volt in this case, which would give R_1 = 910 ohms.

34 The resistor values you calculated in problem 33 ensure that the transistor turns ON and that the 100 mA current (I_C) you need to illuminate the lamp flows through the lamp and the transistor. Figure 4.21 shows the labeled circuit.

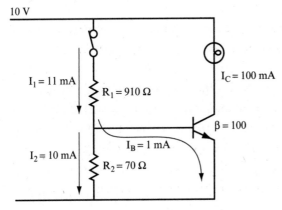

FIGURE 4.21

QUESTIONS

For each of the following lamps, perform the same calculations you used in the last few problems to find the values of R_1 and R_2.

A. A 28-volt lamp that draws 56 mA. $\beta = 100$ _____

_____

B. A 12-volt lamp that draws 140 mA. $\beta = 50$ _____

_____

ANSWERS

A. $I_B = \dfrac{56\,mA}{100} = 0.56\,mA$

$I_2 = 5.6\,mA$

$R_2 = \dfrac{0.7\,volt}{5.6\,mA} = 125\,ohms$

$R_1 = \dfrac{28\,volts}{6.16\,mA} = 4.5k\Omega$

B. $R_2 = 25\,ohms$

$R_1 = 400\,ohms$

35 The arbitrary decision to make the value of I_2 10 times the value of I_B is obviously subject to considerable discussion, doubt, and disagreement. *Transistors are not exact devices; they are not carbon copies of each other.*

In general, any transistor of the same type has a different β from any other because of the variance in tolerances found in component manufacturing. This leads to a degree of inexactness in designing and analyzing transistor circuits. The truth is that if you follow exact mathematical procedures, it can complicate your life. In practice, a few "rules of thumb" have been developed to help you make the necessary assumptions. These rules lead to simple equations that provide workable values for components that you can use in designing circuits.

The choice of $I_2 = 10I_B$ is one such rule of thumb. Is it the only choice that works? Of course not. Almost any value of I_2 that is at least 5 times larger than I_B can work. Choosing 10 times the value is a good option for three reasons:

- It is a good practical choice. It always works.
- It makes the arithmetic easy.
- It's not overly complicated and doesn't involve unnecessary calculations.

QUESTION

In the example from problem 32, $I_B = 1$ mA and $I_2 = 10$ mA. Which of the following values can also work efficiently for I_2?

A. 5 mA

B. 8 mA

C. 175 mA

D. 6.738 mA

E. 1 mA

ANSWER

Choices A, B, and D. Value C is too high to be a sensible choice, and E is too low.

36 Before you continue with this chapter, answer the following review questions.

QUESTIONS

A. Which switches faster, the transistor or the mechanical switch? _____

_____

B. Which can be more accurately controlled? _____

_____

C. Which is the easiest to operate remotely? _____

_____

D. Which is the most reliable? _____

_____

E. Which has the longest life? _____

_____

ANSWERS

A. The transistor is much faster.

B. The transistor.

C. The transistor.

D. The transistor.

E. Because transistors have no moving parts, they have a much longer operating lifetime than a mechanical switch. A mechanical switch will fail after several thousand operations, whereas transistors can be operated several million times a second and can last for years.

SWITCHING THE JFET

37 The use of the junction field effect transistor (JFET) as a switch is discussed in the next few problems. You may want to review problems 28 through 31 in Chapter 3 where this book introduced the JFET.

The JFET is considered a "normally on" device, which means that with 0 volts applied to the input terminal (called the *gate*), it is ON, and current can flow through the transistor. When you apply a voltage to the gate, the device conducts less current because the resistance of the drain to the source channel increases. At some point, as the voltage increases, the value of the resistance in the channel becomes so high that the device "cuts off" the flow of current.

QUESTIONS

A. What are the three terminals for a JFET called, and which one controls the operation of the device? _____

_____

B. What turns the JFET ON and OFF? _____

_____

ANSWERS

A. Drain, source, and gate, with the gate acting as the control.

B. When the gate voltage is zero (at the same potential as the source), the JFET is ON. When the gate to source voltage difference is high, the JFET is OFF.

PROJECT 4.2: The JFET

OBJECTIVE

The objective of this project is to determine the drain current that flows when a JFET is fully ON, and the gate voltage needed to fully shut the JFET OFF, using the circuit shown in Figure 4.22.

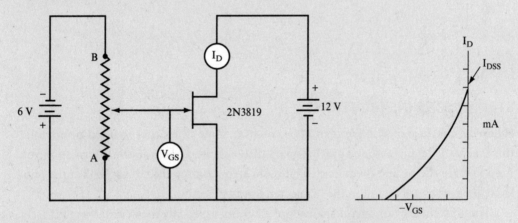

FIGURE 4.22

GENERAL INSTRUCTIONS

After the circuit is set up, change the gate voltage (V_{GS}) by adjusting the potentiometer. Measure the drain current (I_D) for each V_{GS} value. As you work through the project, observe how the drain current drops toward zero as you increase V_{GS}. When the JFET is OFF, I_D is at zero; when the JFET is fully ON, I_D is at its maximum (called I_{DSS}).

Parts List

You need the following equipment and supplies:

- ❏ One 6-volt battery pack (4 AA batteries)
- ❏ One 12-volt battery pack (8 AA batteries)
- ❏ One multimeter set to mA
- ❏ One multimeter set to measure DC voltage
- ❏ One 10 kΩ potentiometer
- ❏ One breadboard
- ❏ Two terminal blocks
- ❏ One 2N3819 JFET (Figure 4.23 shows the pinout for the 2N3819.)

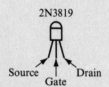

FIGURE 4.23

STEP-BY-STEP INSTRUCTIONS

Set up the circuit shown in Figure 4.22. If you have some experience in building circuits, this schematic (along with the previous parts list) should provide all the information you need to build the circuit. If you need a bit more help building the circuit, look at the photos of the completed circuit in the "Expected Results" section.

Carefully check your circuit against the diagram, especially the orientation of the JFET to ensure that the drain, gate, and source leads are connected correctly. One unusual aspect of this circuit you may want to check is that the +V bus of the 6-volt battery pack should be connected to the ground bus of the 12-volt battery pack.

After you check your circuit, follow these steps, and record your measurements in the blank table following the steps:

1. Adjust the potentiometer to set V_{GS} at 0 volts. (Your multimeter may indicate a few tenths of a millivolt; that's close enough.)
2. Measure and record V_{GS} and I_D.
3. Adjust the potentiometer slightly to give a higher value of V_{GS}.
4. Measure and record the new values of V_{GS} and I_D.

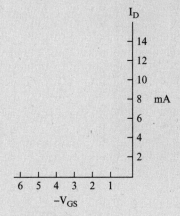

5. Repeat steps 3 and 4 until I_D drops to 0 mA.

V_{GS} (Volts)	I_D (mA)

6. Graph the points recorded in the table, using the blank graph in Figure 4.24. Draw a curve through the points. Your curve should look like the one in Figure 4.22.

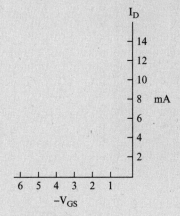

FIGURE 4.24

EXPECTED RESULTS

Figure 4.25 shows the breadboarded circuit for this project.

Drain   Gate   Source

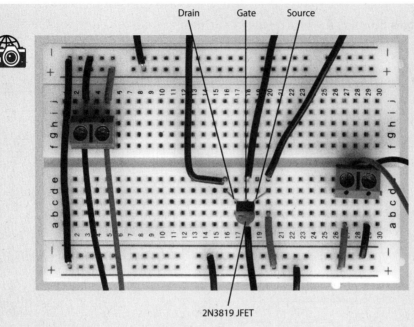

2N3819 JFET

FIGURE 4.25

Figure 4.26 shows the test setup for this project.

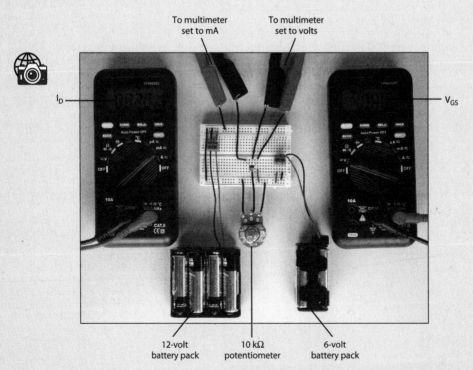

To multimeter set to mA   To multimeter set to volts

I_D V_{GS}

12-volt battery pack   10 kΩ potentiometer   6-volt battery pack

FIGURE 4.26

Compare your measurements with the ones shown in the following table. You should see a similar trend in the measured values, not exactly the same values.

V_{GS} (Volts)	I_D (mA)
0	12.7
0.4	10.7
0.6	9.8
0.8	8.9
1.0	8.1
1.3	6.8
1.5	6.0
1.8	4.9
2.0	4.1
2.3	3.1
2.5	2.5
2.7	1.9
3.0	1.1
3.3	0.5
3.5	0.2
3.7	0.1
4.0	0

Figure 4.27 is the V-I curve generated using the measurements shown in the preceding table. This graph is called the transfer curve for the JFET.

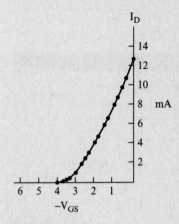

FIGURE 4.27

With the potentiometer set to 0 ohms (point A in Figure 4.22), the voltage from the gate to the source is zero (V_{GS} = 0). The current that flows between the drain and source terminals of the JFET at this time is at its maximum value and is called the *saturation current* (I_{DSS}).

> **NOTE** One property of the saturation current is that when V_{GS} is set at zero, and the transistor is fully ON, the current doesn't drop as long as the value of V_{DS} is above a few volts. If you have an adjustable power supply, you can determine the value of V_{DS} at which I_D starts to drop by starting with the power supply set at 12 volts. Watch the value of I_D as you lower the power supply voltage until you see I_D start to decrease.

38 Refer to the transfer curve shown in Figure 4.28 to answer the following questions.

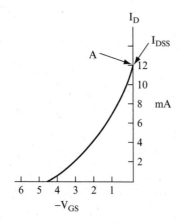

FIGURE 4.28

QUESTIONS

Using the transfer curve shown in Figure 4.28, answer the following:

A. With V_{GS} = 0, what is the value of the drain current? _____

_____

B. Why is this value called the drain saturation current? _____

_____

C. What is the gate to source cutoff voltage for the curve shown? _____

_____

D. Why is this called a cutoff voltage? _____

_____

ANSWERS

A. 12 mA on the graph.

B. The word "saturation" is used to indicate that the current is at its maximum.

C. Approximately –4.5 V on the graph.

D. It is termed a "cutoff voltage" because at this value, the drain current goes to 0 ampere.

39 Now, look at the circuit shown in Figure 4.29. Assume that the JFET has the transfer characteristic shown by the curve in problem 38.

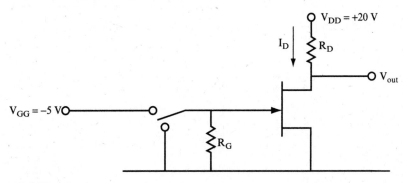

FIGURE 4.29

When the gate is connected to ground, the drain current will be at 12 mA. Assuming that the drain to source resistance is negligible, you can calculate the required value for R_D using the following formula:

$$R_D = \frac{V_{DD}}{I_{DSS}}$$

If you know the drain to source voltage, then you can include it in the calculation.

$$R_D = \frac{(V_{DD} - V_{DS})}{I_{DSS}}$$

QUESTION

What should the value of R_D be for the I_{DSS} shown at point A in the curve? _____

_____

ANSWER

$$R_D = \frac{20\,\text{volts}}{12\,\text{mA}} = 1.67\,\text{k}\Omega$$

40 For the JFET circuit shown in Figure 4.29, assume that $V_{DS} = 1$ volt when the I_D is at saturation.

QUESTIONS

A. What is the required value of R_D? _____

_____

B. What is the effective drain to source resistance (r_{DS}) in this situation? _____

_____

ANSWERS

A. $R_D = \dfrac{(20\,\text{volts} - 1\,\text{volt})}{12\,\text{mA}} = 1583\,\text{ohms}$

B. $r_{DS} = \dfrac{V_{DS}}{I_{DSS}} = \dfrac{1\,\text{volt}}{12\,\text{mA}} = 83\,\text{ohms}$

✳ **NOTE** You can see from this calculation that R_D is 19 times greater than r_{DS}. Thus, ignoring V_{DS} and assuming that $r_{DS} = 0$ does not greatly affect the value of R_D. The 1.67 kΩ value is only about 5 percent higher than the 1583 ohm value for R_D.

41 Now, turn the JFET OFF. From the curve shown in Figure 4.28, you can see that a cutoff value of –4.5 volts is required. Use a gate to source value of –5 volts to ensure that the JFET is in the "hard OFF" state. The purpose of resistor R_G is to ensure that the gate is connected to ground while you flip the switch between terminals, changing the gate voltage from one level to the other. Use a large value of 1 MΩ here to avoid drawing any appreciable current from the gate supply.

QUESTION

When the gate is at the −5 V potential, what is the drain current and the resultant output voltage? _____

_____

ANSWER

I_D = 0 ampere and $V_{out} = V_{DS}$ = 20 volts, which is V_{DD}

SUMMARY

In this chapter, you learned about the transistor switch and how to calculate the resistor values required to use it in a circuit.

- You worked with a lamp as the load example because this provides an easy visual demonstration of the switching action. All the circuits shown in this chapter work when you build them on a breadboard, and the voltage and current measurements are close to those shown in the text.

- You have not yet learned all there is to transistor switching. For example, you haven't found out how much current a transistor can conduct before it burns out, what maximum voltage a transistor can sustain, or how fast a transistor can switch ON and OFF. You can learn these things from the data sheet for each transistor model, so these things are not covered here.

- When you use the JFET as a switch, it does not switch as fast as a BJT, but it does have certain advantages relating to its large input resistance. The JFET does not draw any current from the control circuit to operate. Conversely, a BJT will draw current from the control circuit because of its lower input resistance.

SELF-TEST

These questions test your understanding of the concepts introduced in this chapter. Use a separate sheet of paper for your diagrams or calculations. Compare your answers with the answers provided.

For the first three questions, use the circuit shown in Figure 4.30. The objective is to find the value of R_B that turns the transistor ON. As you may know, resistors are manufactured with "standard values." After you have calculated an exact value, choose the nearest standard resistor value from Appendix D, "Standard Resistor Values."

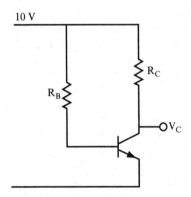

10 V

R_C

R_B

V_C

FIGURE 4.30

1. $R_C = 1\ k\Omega,\ \beta = 100$

   $R_B =$ _____

2. $R_C = 4.7\ k\Omega,\ \beta = 50$

   $R_B =$ _____

3. $R_C = 22\ k\Omega,\ \beta = 75$

   $R_B =$ _____

For questions 4–6, use the circuit shown in Figure 4.31. Find the values of R_3, R_2, and R_1 that ensure that Q_2 is ON or OFF when the switch is in the corresponding position. Calculate the resistors in the order given. After you find the exact values, again choose the nearest standard resistor values.

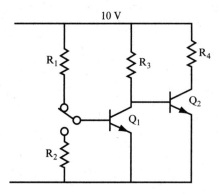

FIGURE 4.31

NOTE   Rounding off throughout a problem, or rounding off the final answer, could produce slightly different results.

4. $R_4 = 100$ ohms, $\beta_1 = 100$, $\beta_2 = 20$.

$R_3 = \rule{6cm}{0.4pt}$

$R_1 = \rule{6cm}{0.4pt}$

$R_2 = \rule{6cm}{0.4pt}$

5. $R_4 = 10$ ohms, $\beta_1 = 50$, $\beta_2 = 20$.

$R_3 = \rule{6cm}{0.4pt}$

$R_1 = \rule{6cm}{0.4pt}$

$R_2 = \rule{6cm}{0.4pt}$

6. $R_4 = 250$ ohms, $\beta_1 = 75$, $\beta_2 = 75$.

$R_3 = \rule{6cm}{0.4pt}$

$R_1 = \rule{6cm}{0.4pt}$

$R_2 = \rule{6cm}{0.4pt}$

For questions 7–9, find the values of the resistors in the circuit shown in Figure 4.32 that ensure that Q_3 will be ON or OFF when the switch is in the corresponding position. Then, select the nearest standard resistor values.

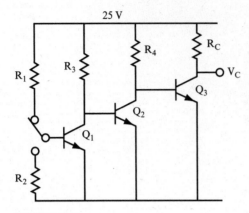

FIGURE 4.32

7. $R_C = 10$ ohms, $\beta_3 = 20$, $\beta_2 = 50$, $\beta_1 = 100$.

$R_4 = \underline{\hspace{10cm}}$

$R_2 = \underline{\hspace{10cm}}$

$R_3 = \underline{\hspace{10cm}}$

$R_1 = \underline{\hspace{10cm}}$

8. $R_C = 28$ ohms, $\beta_3 = 10$, $\beta_2 = 75$, $\beta_1 = 75$.

$R_4 = \underline{\hspace{10cm}}$

$R_2 = \underline{\hspace{10cm}}$

$R_3 = \underline{\hspace{10cm}}$

$R_1 = \underline{\hspace{10cm}}$

9. $R_C = 1$ ohm, $\beta_3 = 10$, $\beta_2 = 50$, $\beta_1 = 75$.

$R_4 = \underline{\hspace{10cm}}$

$R_2 = \underline{\hspace{10cm}}$

$R_3 = \underline{\hspace{10cm}}$

$R_1 = \underline{\hspace{10cm}}$

Questions 10–12 use the circuit shown in Figure 4.33. Find values for R_1 and R_2 that ensure that the transistor turns ON when the switch is closed and OFF when the switch is open.

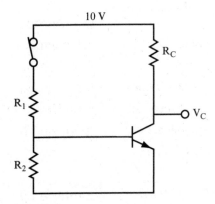

FIGURE 4.33

10. $R_C = 1\ k\Omega,\ \beta = 100.$

$R_1 =$ _____

$R_2 =$ _____

11. $R_C = 22\ k\Omega,\ \beta = 75.$

$R_1 =$ _____

$R_2 =$ _____

12. $R_C = 100\ \Omega,\ \beta = 30.$

$R_1 =$ _____

$R_2 =$ _____

13. An N-channel JFET has a transfer curve with the following characteristics. When $V_{GS} = 0$ volt, the saturation current (I_{DSS}) is 10.5 mA, and the cutoff voltage is –3.8 volts. With a drain supply of 20 volts, design a biasing circuit that switches the JFET from the ON state to the OFF state.

ANSWERS TO SELF-TEST

The exercises in this Self-Test show calculations that are typical of those found in practice, and the odd results you sometimes get are quite common. Thus, choosing a nearest standard value of resistor is a common practice. If your answers do not agree with those given here, review the problems indicated in parentheses before you go on to Chapter 5.

1.	100 kΩ	(problem 8)
2.	235 kΩ. Choose 240 kΩ as a standard value.	(problem 8)
3.	1.65 MΩ. Choose 1.6 MΩ as a standard value.	(problem 8)
4.	$R_3 = 2$ kΩ; $R_1 = 200$ kΩ; $R_2 = 200$ kΩ. Use these values.	(problem 22)
5.	$R_3 = 200$ ohms; $R_1 = 10$ kΩ; $R_2 = 10$ kΩ. Use these values.	(problem 22)
6.	$R_3 = 18.8$ kΩ. Choose 18 kΩ as a standard value.	(problem 22)
	$R_1 = 1.41$ MΩ. Choose 1.5 MΩ as a standard value.	
	Select 1 MΩ for R_2.	
7.	$R_4 = 200$ ohms; $R_3 = 10$ kΩ; $R_2 = 1$ MΩ; $R_1 = 1$ MΩ. Use these values.	(problem 26)
8.	$R_4 = 280$ ohms. Choose 270 ohms as a standard value.	(problem 26)
	$R_3 = 21$ kΩ. Choose 22 kΩ as a standard value.	
	$R_2 = 1.56$ MΩ. Choose 1.5 or 1.6 MΩ as a standard value.	
	$R_1 = 1.56$ MΩ. Choose 1.5 or 1.6 MΩ as a standard value.	
9.	$R_4 = 10$ ohms. Choose 10 ohms as a standard value.	(problem 26)
	$R_3 = 500$ ohms. Choose 510 ohms as a standard value.	
	$R_2 = 37.5$ kΩ. Choose 39 kΩ as a standard value.	
	$R_1 = 37.5$ kΩ. Choose 39 kΩ as a standard value.	
10.	$R_2 = 700$ ohms. Choose 680 or 720 ohms as a standard value.	(problems 31–33)
	$R_1 = 8.45$ kΩ. Choose 8.2 kΩ as a standard value.	
	If 0.7 is ignored, then $R_1 = 9.1$ kΩ.	
11.	$R_2 = 11.7$ kΩ. Choose 12 KΩ as a standard value.	(problems 31–33)
	$R_1 = 141$ kΩ. Choose 140 or 150 kΩ as a standard value.	
12.	$R_2 = 21$ ohms. Choose 22 ohms as a standard value.	(problems 31–33)
	$R_1 = 273$ ohms. Choose 270 ohms as a standard value.	
13.	Use the circuit shown in Figure 4.29. Set the gate supply at a value slightly more negative than –3.8 volts. A value of –4 V would work. Make resistor $R_G = 1$ MΩ. Set R_D at a value of (20 volts)/(10.5 mA), which calculates a resistance of 1.9 kΩ. You can wire a standard resistor of 1 kΩ in series with a standard resistor of 910 ohms to obtain a resistance of 1.91 kΩ.	(problems 39 and 41)

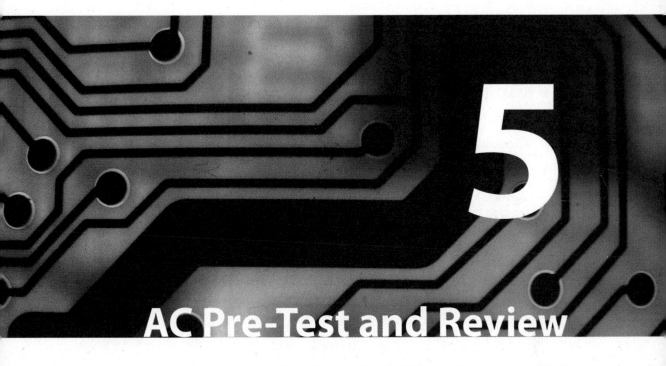

AC Pre-Test and Review

You need to have some basic knowledge of alternating current (AC) to study electronics. To understand AC, you must understand sine waves.

A *sine wave* is simply a shape, like waves in the ocean. Sine waves in electronics are used to represent voltage or current moving up and down in magnitude. In AC electronics, some signals or power sources (such as the house current provided at a wall plug) are represented by sine waves. The sine wave shows how the voltage moves from 0 volts to its peak voltage, and back down through 0 volts at 60 cycles per second, or 60 Hertz (Hz).

The sound from a musical instrument also consists of sine waves. When you combine sounds (such as all the instruments in an orchestra), you get complex combinations of many sine waves at various frequencies.

The study of AC starts with the properties of simple sine waves and continues with an examination of how electronic circuits can generate or change sine waves.

This chapter discusses the following:

- Generators
- Sine waves
- Peak-to-peak and root mean square voltages
- Resistors in AC circuits
- Capacitive and inductive reactance
- Resonance

THE GENERATOR

1 In electronic circuits powered by direct current (DC), the voltage source is usually a battery or solar cell, which produces a constant voltage and a constant current through a conductor.

In electronic circuits or devices powered by alternating current (AC), the voltage source is usually a *generator*, which produces a regular output waveform, such as a sine wave.

QUESTION

Draw one cycle of a sine wave.

ANSWER

Figure 5.1 shows one cycle of a sine wave.

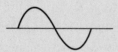

FIGURE 5.1

2 A number of electronic instruments are used in the laboratory to produce sine waves. For purposes of this discussion, the term *generator* means a sine wave source. These generators enable you to adjust the voltage and frequency by turning a dial or pushing a button. These instruments are called by various names, generally based on the method of producing the sine wave, or the application as a test instrument. The most popular generator at present is called a *function generator*. It provides a choice of functions or waveforms, including a square wave and a triangle wave. These waveforms are useful in testing certain electronic circuits.

The symbol shown in Figure 5.2 represents a generator. Note that a sine wave shown within a circle designates an AC sine wave source.

FIGURE 5.2

QUESTIONS

A. What is the most popular instrument used in the lab to produce waveforms? _____
_____

B. What does the term AC mean? _____
_____

C. What does the sine wave inside a generator symbol indicate? _____
_____

ANSWERS

A. Function generator.

B. Alternating current, as opposed to direct current.

C. The generator is a sine wave source.

3 Figure 5.3 shows some key parameters of sine waves . The two axes are voltage and time.

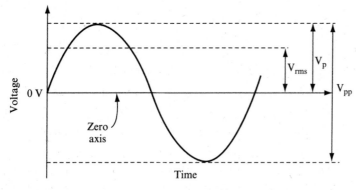

FIGURE 5.3

The *zero axis* is the reference point from which all voltage measurements are made.

QUESTIONS

A. What is the purpose of the zero axis? _____

_____

B. What is the usual point for making time measurements? _____

_____

ANSWERS

A. It is the reference point from which all voltage measurements are made.

B. Time measurements can be made from any point in the sine wave, but usually they are made from a point at which the sine wave crosses the zero axis.

4 The three most important voltage or amplitude measurements are the *peak (p), peak-to-peak (pp),* and the *root mean square (rms)* voltages.

The following equations show the relationship between p, pp, and rms voltages for sine waves. The relationships between p, pp, and rms voltages differ for other waveforms (such as square waves).

$$V_p a = \sqrt{2} \times V_{rms}$$

$$V_{pp} = 2V_p = 2 \times \sqrt{2} \times V_{rms}$$

$$V_{rms} = \frac{1}{\sqrt{2}} \times V_p = \frac{1}{\sqrt{2}} \times \frac{V_{pp}}{2}$$

Note the following:

$$\sqrt{2} = 1.414$$

$$\frac{1}{\sqrt{2}} = 0.707$$

QUESTION

If the pp voltage of a sine wave is 10 volts, find the rms voltage. _____

ANSWER

$$V_{rms} = \frac{1}{\sqrt{2}} \times \frac{V_{pp}}{2} = 0.707 \times \frac{10}{2} = 3.535 \, V$$

5 Calculate the following for a sine wave.

QUESTION

If the rms voltage is 2 volts, find the pp voltage. _____

ANSWER

$$V_{pp} = 2 \times \sqrt{2} \times V_{rms} = 2 \times 1.414 \times 2 = 5.656 \, V$$

6 Calculate the following for a sine wave.

QUESTIONS

A. V_{pp} = 220 volts. Find V_{rms}. _____

B. V_{rms} = 120 volts. Find V_{pp}. _____

ANSWERS

A. 77.77 volts

B. 340 volts (This is the common house current supply voltage; 340 V_{pp} = 120 V_{rms}.)

7 There is a primary time measurement for sine waves. The duration of the complete sine wave is shown in Figure 5.4 and referred to as a *cycle*. All other time measurements are fractions or multiples of a cycle.

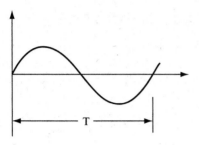

FIGURE 5.4

QUESTIONS

A. What is one complete sine wave called? _____

_____

B. What do you call the time it takes to complete one sine wave? _____

_____

C. How is the frequency of a sine wave related to this time? _____

_____

D. What is the unit for frequency? _____

_____

E. If the period of a sine wave is 0.5 ms, what is its frequency? What is the frequency of a sine wave with a period of 40 μsec? _____

_____

F. If the frequency of a sine wave is 60 Hz, what is its period? What is the period of sine waves with frequencies of 12.5 kHz and 1 MHz? _____

_____

ANSWERS

A. Cycle

B. The period, T

8 Choose all answers that apply.

QUESTIONS

Which of the following could represent electrical AC signals?

A. Simple sine wave

B. Mixture of many sine waves, of different frequencies and amplitudes

C. Straight line

ANSWERS

A and B

RESISTORS IN AC CIRCUITS

9 Alternating current is passed through components, just as direct current is. Resistors interact with alternating current just as they do with direct current.

QUESTION

Suppose an AC signal of 10 V_{pp} is connected across a 10-ohm resistor. What is the current through the resistor? _____

_____

ANSWER

Use Ohm's law:

$$I = \frac{V}{R} = \frac{10\,V_{pp}}{10\,ohms} = 1\,A_{pp}$$

Because the voltage is given in pp, the current is a pp current.

10 An AC signal of 10 V_{rms} is connected across a 20-ohm resistor.

QUESTION

Find the current. _____

_____

ANSWER

$$I = \frac{10\,V_{rms}}{20\,ohms} = 0.5\,A_{rms}$$

Because the voltage was given in rms, the current is in rms.

11 You apply an AC signal of 10 V_{pp} to the voltage divider circuit, as shown in Figure 5.5.

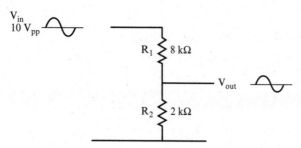

V_{in}
10 V_{pp}

R_1 8 kΩ

V_{out}

R_2 2 kΩ

FIGURE 5.5

QUESTION

Find V_{out}. _____

_____

ANSWER

$$V_{out} = V_{in} \times \frac{R_2}{(R_1 + R_2)} = 10 \times \frac{2k\Omega}{(8k\Omega + 2k\Omega)} = 10 \times \frac{2}{10} = 2V_{pp}$$

CAPACITORS IN AC CIRCUITS

12 A capacitor opposes the flow of an AC current.

QUESTIONS

A. What is this opposition to the current flow called? _____

_____

B. What is this similar to in DC circuits? _____

_____

ANSWERS

A. Reactance

B. Resistance

13 Just as with resistance, you determine reactance by using an equation.

QUESTIONS

A. What is the equation for reactance? _____

_____

B. What does each symbol in the equation stand for? _____

_____

C. How does the reactance of a capacitor change as the frequency of a signal increases? _____

_____

ANSWERS

A. $X_C = \dfrac{1}{2\pi fC}$

B. X_C = the reactance of the capacitor in ohms.

f = the frequency of the signal in hertz.

C = the value of the capacitor in farads.

C. The reactance of a capacitor decreases as the frequency of the signal increases.

14 Assume the capacitance is 1 µF and the frequency is 1 kHz.

QUESTION

Find the capacitor's reactance. (*Note:* 1/(2π) = 0.159, approximately.) _____

_____

ANSWER

$X_C = \dfrac{1}{2\pi fC}$

f = 1 kHz = 10^3 Hz

C = 1 µF = 10^{-6} F

Thus,

$X_C = \dfrac{0.159}{10^3 \times 10^{-6}} = 160\,\text{ohms}$

15 Now, perform these simple calculations. In each case, find X_{C1} (the capacitor's reactance at 1 kHz) and X_{C2} (the capacitor's reactance at the frequency specified in the question).

QUESTIONS

Find X_{C1} and X_{C2}:

A. $C = 0.1\ \mu F$, $f = 100$ Hz. _____

_____

B. $C = 100\ \mu F$, $f = 2$ kHz. _____

_____

ANSWERS

A. At 1 kHz, $X_{C1} = 1600$ ohms; at 100 Hz, $X_{C2} = 16,000$ ohms

B. At 1 kHz, $X_{C1} = 1.6$ ohms; at 2 kHz, $X_{C2} = 0.8$ ohms

A circuit containing a capacitor in series with a resistor (as shown in Figure 5.6) functions as a voltage divider.

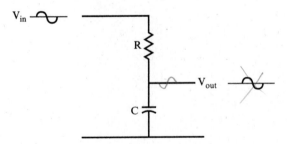

FIGURE 5.6

Although this voltage divider provides a reduced output voltage, just like a voltage divider using two resistors, there's a complication. If you view the output and input voltage waveforms on an oscilloscope, you see that one is shifted away from the other. The two waveforms are said to be "out of phase." *Phase* is an important concept in understanding how certain electronic circuits work. In Chapter 6, "Filters," you learn about phase relationships for some AC circuits. You also encounter this again when you study amplifiers.

USING THE OSCILLOSCOPE

You use an oscilloscope to measure AC signals generated by a circuit, or to measure the effect that a circuit has on AC signals. The key parameters you measure with an oscilloscope are *frequency* and *peak-to-peak voltage*. An oscilloscope can also be used to show the shape of a signal's waveform so that you can ensure that the circuit works properly. When using an oscilloscope to compare a circuit's input signal to its output signal, you can determine the phase shift, as well as the change in V_{pp}.

The following figure shows an oscilloscope whose probe connects to the output of an oscillator circuit to measure the frequency of the signal generated by the oscillator. (Oscillator circuits are discussed in Chapter 9, "Oscillators.") This example uses an analog oscilloscope, but you can also use a digital oscilloscope, which automates many of the measurements.

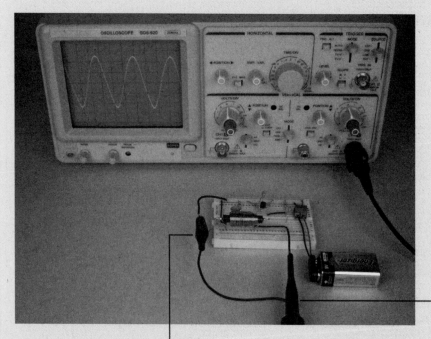

Oscilloscope probe

Ground clip

This oscilloscope has two channels, which provide the capability to measure two waveforms at once. Only channel 2 was used for the measurement in the preceding figure. The oscilloscope probe was clipped to a jumper wire connecting to

V_{out} for the circuit, and the oscilloscope ground clip was clipped to a jumper wire connecting to the ground bus.

The following figure shows the oscilloscope control panel. You use the VOLTS/DIV control to adjust the vertical scale and the TIME/DIV control to adjust the horizontal scale. Set the vertical position knob and the horizontal position knob to adjust the position of the waveform against the grid to make it easier to measure.

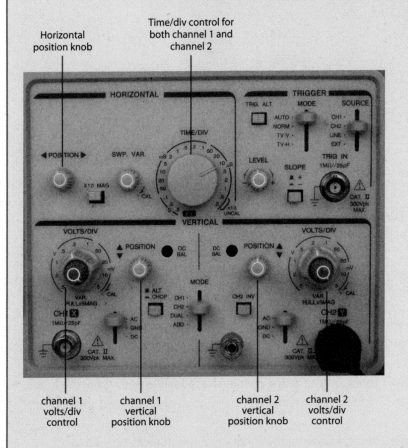

You can determine the period of this waveform by counting the number of horizontal divisions the waveform takes to complete one cycle, and then multiplying the number of divisions by the TIME/DIV setting. In the following figure, the period of the sine wave generated by this oscillator circuit is approximately 3.3 divisions wide.

Continued

(continued)

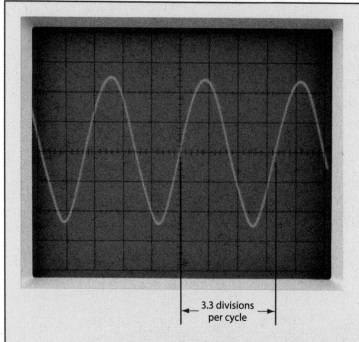

|←——— 3.3 divisions ———→|
per cycle

Because the TIME/DIV knob is set at 10 μs, the period of this sine wave is 33 μs. The frequency of this sine wave is therefore calculated as follows:

$$f = \frac{1}{T} = \frac{1}{33\mu s} = \frac{1}{.000033\,sec} = 30303 \text{ Hz} = 30.3 \text{ kHz}$$

You can also measure the effect of a circuit on a signal of a particular frequency. Supply the signal from a function generator to the input of the circuit. Attach the oscilloscope probe for channel 2 to the input of the circuit. Attach the oscilloscope probe for channel 1 to the output of the circuit. The following figure shows a function generator and oscilloscope attached to the voltage divider circuit shown in Figure 5.6.

In this example, the red lead from the function generator was clipped to a jumper wire connected to the resistor in the voltage divider circuit, and the black lead was clipped to a jumper wire connected to the ground bus. The oscilloscope probe for channel 2 is clipped to a jumper wire connected to the resistor, and the ground clip is clipped to a jumper wire attached to the ground bus. The oscilloscope probe for channel 1 is clipped to a jumper wire connected to the voltage divider circuit V_{out}, and the ground clip is clipped to a jumper wire connected to the ground bus.

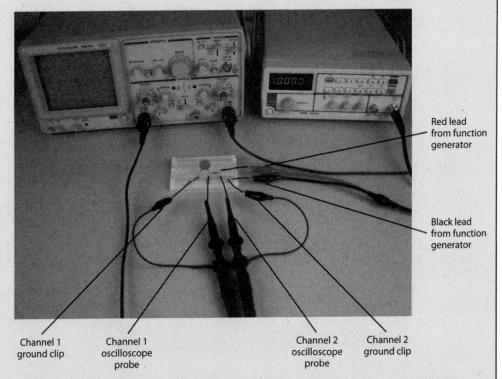

Red lead from function generator

Black lead from function generator

Channel 1 ground clip

Channel 1 oscilloscope probe

Channel 2 oscilloscope probe

Channel 2 ground clip

The function generator supplies an input signal at a frequency of 10 kHz and an amplitude of 10 V_{pp}. The input signal is represented by the upper sine wave on the oscilloscope. Many function generators (such as the one shown here) have an amplitude adjustment knob without a readout. You set the input signal amplitude to 10 V_{pp} with the amplitude knob on the function generator while monitoring the amplitude on the oscilloscope.

The output signal is represented by the lower sine wave in the following figure. Adjust the VOLT/DIV controls and vertical position controls for channels 1 and 2 to fit both sine waves on the screen, as shown here.

You can measure V_{pp} for each sine wave by multiplying the number of vertical divisions between peaks by the setting on the VOLT/DIV knobs. For the input sine wave in this example, this measurement is two divisions at 5 VOLTS/DIV,

Continued

(continued)

for a total of 10 volts. For the output sine wave, this measurement is 3 divisions at 2 VOLTS/DIV, for a total of 6 volts. This indicates that the circuit has decreased the input signal from 10 V$_{pp}$ to 6 V$_{pp}$.

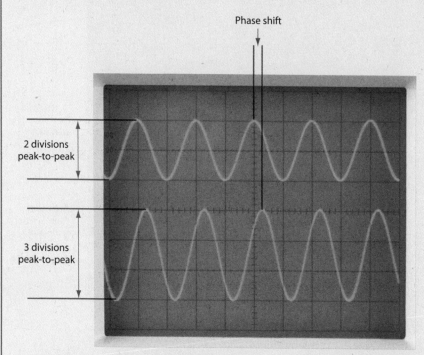

Also note that that the peak of the output waveform shifts from the input waveform, a phenomenon called *phase shift*. You learn more about how to calculate phase shift in Chapter 6.

THE INDUCTOR IN AN AC CIRCUIT

16 An *inductor* is a coil of wire, usually wound many times around a piece of soft iron. In some cases, the wire is wound around a nonconducting material.

QUESTIONS

A. Is the AC reactance of an inductor high or low? Why? _____

_____

B. Is the DC resistance high or low? _____

_____

C. What is the relationship between the AC reactance and the DC resistance? _____

_____

D. What is the formula for the reactance of an inductor? _____

_____

ANSWERS

A. Its AC reactance (X_L), which can be quite high, is a result of the electromagnetic field that surrounds the coil and induces a current in the opposite direction of the original current.

B. Its DC resistance (r), which is usually quite low, is simply the resistance of the wire that makes up the coil.

C. None

D. $X_L = 2\pi fL$, where L = the value of the inductance in henrys (H). Using this equation, you can expect the reactance of an inductor to increase as the frequency of a signal passing through it increases.

17 Assume the inductance value is 10 henrys (H) and the frequency is 100 Hz.

QUESTION

Find the reactance. _____

_____

ANSWER

$X_L = 2\pi fL = 2\pi \times 100 \times 10 = 6280$ ohms

18 Now, try these two examples. In each case, find X_{L1} (the reactance of the inductor at 1 kHz) and X_{L2} (the reactance at the frequency given in the question).

QUESTIONS

A. $L = 1$ mH (0.001 H), $f = 10$ kHz _____

_____

B. $L = 0.01$ mH, $f = 5$ MHz _____

_____

ANSWERS

A. $X_{L1} = 6.28 \times 10^3 \times 0.001 = 6.28$ ohms

$X_{L2} = 6.28 \times 10 \times 10^3 \times 0.001 = 62.8$ ohms

B. $X_{L1} = 6.28 \times 10^3 \times 0.01 \times 10^{-3} = 0.0628$ ohms

$X_{L2} = 6.28 \times 5 \times 10^6 \times 0.01 \times 10^{-3} = 314$ ohms

A circuit containing an inductor in series with a resistor functions as a voltage divider, just as a circuit containing a capacitor in series with a resistor does. Again, the relationship between the input and output voltages is not as simple as a resistive divider. The circuit is discussed in Chapter 6.

RESONANCE

19 Calculations in previous problems demonstrate that capacitive reactance decreases as frequency increases, and that inductive reactance increases as frequency increases. If a capacitor and an inductor are connected in series, there is one frequency at which their reactance values are equal.

QUESTIONS

A. What is this frequency called? _____

_____

B. What is the formula for calculating this frequency? You can find it by setting $X_L = X_C$ and solving for frequency. _____

_____

ANSWERS

A. The resonant frequency

B. $2\pi fL = 1/(2\pi fC)$. Rearranging the terms in this equation to solve for f yields the following formula for the resonant frequency (f_r):

$$f_r = \frac{1}{2\pi\sqrt{LC}}$$

20 If a capacitor and an inductor are connected in parallel, there is also a resonant frequency. Analysis of a parallel resonant circuit is not as simple as it is for a series resonant circuit. The reason for this is that inductors always have some internal resistance, which complicates some of the equations. However, under certain conditions, the analysis is similar. For example, if the reactance of the inductor in ohms is more than 10 times greater than its own internal resistance (r), the formula for the resonant frequency is the same as if the inductor and capacitor were connected in series. This is an approximation that you use often.

QUESTIONS

For the following inductors, determine if the reactance is more or less than 10 times its internal resistance. A resonant frequency is provided.

A. f_r = 25 kHz, L = 2 mH, r = 20 ohms _____

_____

B. f_r = 1 kHz, L = 33.5 mH, r = 30 ohms _____

_____

ANSWERS

A. X_L = 314 ohms, which is more than 10 times greater than r

B. X_L = 210 ohms, which is less than 10 times greater than r

NOTE Chapter 7, "Resonant Circuits," discusses both series and parallel resonant circuits. At that time, you learn many useful techniques and formulas.

21 Find the resonant frequency (f_r) for the following capacitors and inductors when they are connected both in parallel and in series. Assume r is negligible.

QUESTIONS

Determine f_r for the following:

A. C = 1 μF, L = 1 henry _____

_____

B. C = 0.2 μF, L = 3.3 mH _____

_____

ANSWERS

A. $f_r = \dfrac{0.159}{\sqrt{10^{-6} \times 1}} = 160\,\text{Hz}$

B. $f_r = \dfrac{0.159}{\sqrt{3.3 \times 10^{-3} \times 0.2 \times 10^{-6}}} = 6.2\,\text{kHz}$

22 Now, try these two final examples.

QUESTIONS

Determine f_r:

A. C = 10 μF, L = 1 henry _____

_____

B. C = 0.0033 μF, L = 0.5 mH _____

_____

ANSWERS

A. f_r = 50 Hz (approximately)

B. f_r = 124 kHz

Understanding resonance is important to understanding certain electronic circuits, such as filters and oscillators.

Filters are electronic circuits that either block a certain band of frequencies, or pass a certain band of frequencies. One common use of filters is in circuits used for radio, television, and other communications applications. *Oscillators* are electronic circuits that generate a continuous output without an input signal. The type of oscillator that uses a resonant circuit produces pure sine waves. (You learn more about oscillators in Chapter 9.)

SUMMARY

Following are the concepts presented in this chapter:

- The sine wave is used extensively in AC circuits.
- The most common laboratory generator is the function generator.
- $V_p = \sqrt{2} \times V_{rms}, V_{pp} = 2\sqrt{2} = V_{rms}$
- $f = 1/T$
- $I_{pp} = \dfrac{V_{pp}}{R}, I_{rms} = \dfrac{V_{rms}}{R}$
- Capacitive reactance is calculated as follows:

$$X_C = \frac{1}{(2\pi fC)}$$

- Inductive reactance is calculated as follows:

$$X_L = 2\pi fL$$

- Resonant frequency is calculated as follows:

$$f_r = \frac{1}{2\pi\sqrt{LC}}$$

SELF-TEST

The following problems test your understanding of the basic concepts presented in this chapter. Use a separate sheet of paper for calculations if necessary. Compare your answers with the answers provided following the test.

1. Convert the following peak or peak-to-peak values to rms values:

    A. $V_p = 12$ V

    $V_{rms} = $ _____

B. $V_p = 80$ mV

$V_{rms} = $ _____

C. $V_{pp} = 100$ V

$V_{rms} = $ _____

2. Convert the following rms values to the required values shown:

A. $V_{rms} = 120$ V

$V_p = $ _____

B. $V_{rms} = 100$ mV

$V_p = $ _____

C. $V_{rms} = 12$ V

$V_{pp} = $ _____

3. For the given value, find the period or frequency:

A. T = 16.7 ms

$f = $ _____

B. f = 15 kHz

$T = $ _____

4. For the circuit shown in Figure 5.7, find the total current flow and the voltage across R_2, (V_{out}). _____

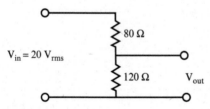

$V_{in} = 20$ V$_{rms}$ 80 Ω

120 Ω V_{out}

FIGURE 5.7

5. Find the reactance of the following components:

A. C = 0.16 μF, f = 12 kHz

$X_C = $ _____

B. L = 5 mH, f = 30 kHz

$X_L = $ _____

6. Find the frequency necessary to cause each reactance shown:

    A. $C = 1 \, \mu F$, $X_C = 200$ ohms

        f = _____

    B. $L = 50 \, mH$, $X_L = 320$ ohms

        f = _____

7. What would be the resonant frequency for the capacitor and inductor values given in A and B of question 5 if they were connected in series? _____

8. What would be the resonant frequency for the capacitor and inductor values given in A and B of question 6 if they were connected in parallel? What assumption would you need to make? _____

ANSWERS TO SELF-TEST

If your answers do not agree with those provided here, review the problems indicated in parentheses before you go to Chapter 6, "Filters."

1A.	$8.5 \, V_{rms}$	(problems 4–6)
1B.	$56.6 \, _mV_{rms}$	
1C.	$35.4 \, V_{rms}$	
2A.	$169.7 \, V_p$	(problems 4–6)
2B.	$141.4 \, mV_p$	
2C.	$33.9 \, V_{pp}$	
3A.	60 Hz	(problem 7)
3B.	66.7 μsec	
4.	$IT = 0.1A_{rms}$, $V_{out} = 12V_{rms}$	(problems 9–11)
5A.	82.9 ohms	(problems 14 and 17)
5B.	942.5 ohms	
6A.	795.8 Hz	(problems 14 and 17)
6B.	1.02 kHz	
7.	5.63 kHz	(problem 19)
8.	711.8 Hz. Assume the internal resistance of the inductor is negligible.	(problem 20)

6

Filters

Certain types of circuits are found in most electronic devices used to process alternating current (AC) signals. One of the most common of these, *filter circuits*, is covered in this chapter. Filter circuits are formed by resistors and capacitors (RC), or resistors and inductors (RL). These circuits (and their effect on AC signals) play a major part in communications, consumer electronics, and industrial controls.

When you complete this chapter, you will be able to do the following:

- Calculate the output voltage of an AC signal after it passes through a high-pass RC filter circuit.

- Calculate the output voltage of an AC signal after it passes through a low-pass RC circuit.

- Calculate the output voltage of an AC signal after it passes through a high-pass RL circuit.

- Calculate the output voltage of an AC signal after it passes through a low-pass RL circuit.

- Draw the output waveform of an AC or combined AC-DC signal after it passes through a filter circuit.

- Calculate simple phase angles and phase differences.

CAPACITORS IN AC CIRCUITS

1 An AC signal is continually changing, whether it is a pure sine wave or a complex signal made up of many sine waves. If such a signal is applied to one plate of a capacitor, it will be induced on the other plate. To express this another way, a capacitor will "pass" an AC signal, as illustrated in Figure 6.1.

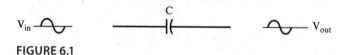

V_{in} ∿ ⎯⎯⎯⎯⎯ C ⊣⊢ ⎯⎯⎯⎯⎯ ∿ V_{out}

FIGURE 6.1

> **NOTE** Unlike an AC signal, a DC signal is blocked by a capacitor. Equally important is that a capacitor is *not* a short circuit to an AC signal.

QUESTIONS

A. What is the main difference in the effect of a capacitor upon an AC signal versus a DC signal? _____

_____

B. Does a capacitor appear as a short or an open circuit to an AC signal? _____

_____

ANSWERS

A. A capacitor will pass an AC signal, whereas it will not pass a DC voltage level.

B. Neither.

2 In general, a capacitor will oppose the flow of an AC current to some degree. As you saw in Chapter 5, "AC Pre-Test and Review," this opposition to current flow is called the *reactance* of the capacitor.

Reactance is similar to resistance, except that the reactance of a capacitor changes when you vary the frequency of a signal. The reactance of a capacitor can be calculated by a formula introduced in Chapter 5.

QUESTION

Write the formula for the reactance of a capacitor. _____

_____

ANSWER

$$X_C = \frac{1}{2\pi fC}$$

3 From this formula, you can see that the reactance changes when the frequency of the input signal changes.

QUESTION

If the frequency increases, what happens to the reactance? _____

_____

ANSWER

It decreases.

If you had difficulty with these first three problems, you should review the examples in Chapter 5.

CAPACITORS AND RESISTORS IN SERIES

4 For simplicity, consider all inputs at this time to be pure sine waves. The circuit shown in Figure 6.2 shows a sine wave as the input signal to a capacitor.

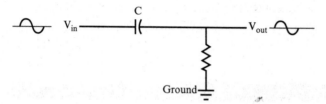

FIGURE 6.2

QUESTION

If the input is a pure sine wave, what is the output? _____

_____

ANSWER

A pure sine wave

---

5 The output sine wave has the same frequency as the input sine wave. A capacitor cannot change the frequency of the signal. But remember, with an AC input, the capacitor behaves in a manner similar to a resistor in that the capacitor does have some level of opposition to the flow of alternating current. The level of opposition depends upon the value of the capacitor and the frequency of the signal. Therefore, the output amplitude of a sine wave will be less than the input amplitude.

QUESTION

With an AC input to a simple circuit like the one described here, what does the capacitor appear to behave like? _____

_____

ANSWER

It appears to have opposition to alternating current similar to the behavior of a resistor.

6 If you connect a capacitor and resistor in series (as shown in Figure 6.3), the circuit functions as a voltage divider.

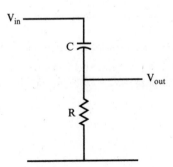

FIGURE 6.3

QUESTION

What formula would you use to calculate the output voltage for a voltage divider formed by connecting two resistors in series? _____

_____

ANSWER

$$V_{out} = V_{in} \times \frac{R_2}{R_1 + R_2}$$

7 You can calculate a total resistance to the flow of electric current for a circuit containing two resistors in series.

QUESTION

What is the formula for this total resistance? _____

_____

ANSWER

$$R_T = R_1 + R_2$$

8 You can also calculate the total opposition to the flow of electric current for a circuit containing a capacitor and resistor in series. This parameter is called *impedance*, and you can calculate it using the following formula:

$$Z = \sqrt{X_C^2 + R^2}$$

In this equation:

$Z =$ The impedance of the circuit in ohms

$X_C =$ The reactance of the capacitor in ohms

$R =$ The resistance of the resistor in ohms

QUESTIONS

Use the following steps to calculate the impedance of the circuit, and the current flowing through the circuit, as shown in Figure 6.4.

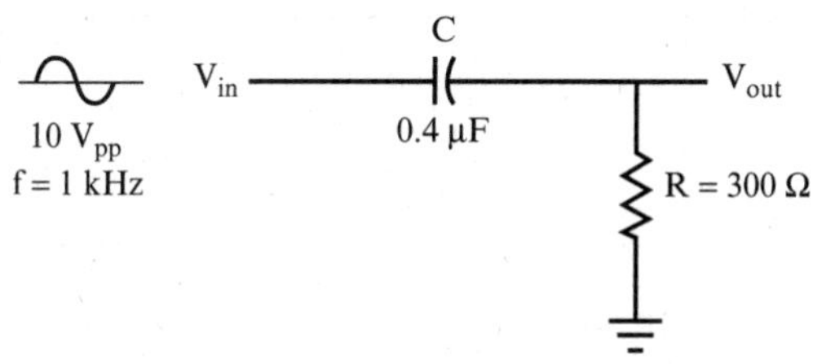

FIGURE 6.4

A. $X_C = \dfrac{1}{2\pi fC} = $ _____

_____

B. $Z = \sqrt{X_C^2 + R^2} = $ _____

_____

C. $I = \dfrac{V}{Z} = $ _____

_____

ANSWERS

A. 400 ohms

B. 500 ohms

C. 20 mA$_{pp}$

9 Now, for the circuit shown in Figure 6.4, calculate the impedance and current using the values provided.

QUESTIONS

A. C = 530 μF, R = 12 ohms, V_{in} = 26 V_{pp}, f = 60 Hz _____

_____

B. C = 1.77 μF, R = 12 ohms, V_{in} = 150 V_{pp}, f = 10 kHz _____

_____

ANSWERS

A. $Z = 13$ ohms, $I = 2$ A_{pp}

B. $Z = 15$ ohms, $I = 10$ A_{pp}

10 You can calculate V_{out} for the circuit shown in Figure 6.5 with a formula similar to the formula used in Chapter 5 to calculate V_{out} for a voltage divider composed of two resistors.

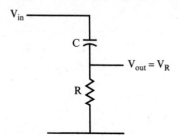

FIGURE 6.5

The formula to calculate the output voltage for this circuit is as follows:

$$V_{out} = V_{in} \times \frac{R}{Z}$$

QUESTIONS

Calculate the output voltage in this circuit using the component values and input signal voltage and frequency listed on the circuit diagram shown in Figure 6.6.

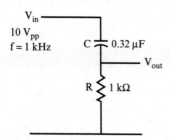

FIGURE 6.6

A. Find X_C: _____

_____

B. Find Z: _____

_____

C. Use the formula to find V_{out}: _____

_____

ANSWERS

A. X_C = 500 ohms (rounded off)

B. Z = 1120 ohms (rounded off)

C. V_{out} = 8.9 V_{pp}

11 Now, find V_{out} for the circuit in Figure 6.5 using the given component values, signal voltage, and frequency.

QUESTIONS

A. C = 0.16 μF, R = 1 kΩ, V_{in} = 10 V_{pp}, f = 1 kHz _____

_____

B. C = 0.08 μF, R = 1 kΩ, V_{in} = 10 V_{pp}, f = 1 kHz _____

_____

ANSWERS

A. V_{out} = 7.1 V_{pp}

B. V_{out} = 4.5 V_{pp}

NOTE Hereafter, you can assume that the answer is a peak-to-peak value if the given value is a peak-to-peak value.

12 The output voltage is said to be *attenuated* in the voltage divider calculations, as shown in the calculations in problems 10 and 11. Compare the input and output voltages in problems 10 and 11.

QUESTION

What does *attenuated* mean? _____

_____

ANSWER

To reduce in amplitude or magnitude (that is, V_{out} is smaller than $V_{in.}$).

13 When you calculated V_{out} in the examples in problems 10 and 11, you first had to find X_C. However, X_C changes as the frequency changes, while the resistance remains constant. Therefore, as the frequency changes, the impedance Z changes and also so does the amplitude of the output voltage V_{out}.

If V_{out} is plotted against frequency on a graph, the curve looks like that shown in Figure 6.7.

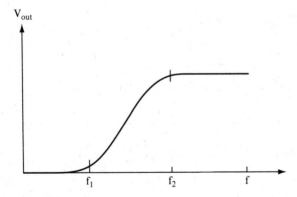

FIGURE 6.7

The frequencies of f_1 (at which the curve starts to rise) and f_2 (where it starts to level off) depend on the values of the capacitor and the resistor.

QUESTIONS

Calculate the output voltage for the circuit shown in Figure 6.8 for frequencies of 100 Hz, 1 kHz, 10 kHz, and 100 kHz.

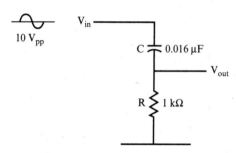

FIGURE 6.8

A. 100 Hz: _____

_____

B. 1 kHz: _____

_____

C. 10 kHz: _____

_____

D. 100 kHz: _____

_____

E. Plot these values for V_{out} against f, and draw a curve to fit the points. Use a separate sheet of paper to draw your graph.

ANSWERS

A. V_{out} = 0.1 volt

B. V_{out} = 1 volt

C. V_{out} = 7.1 volts

D. V_{out} = 10 volts

E. The curve is shown in Figure 6.9.

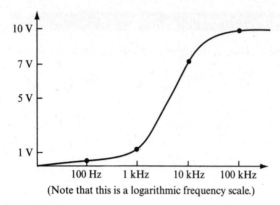

(Note that this is a logarithmic frequency scale.)

FIGURE 6.9

> **NOTE** You can see that V_{out} is equal to V_{in} for the highest frequency and at nearly zero for the lowest frequency. You call this type of circuit a *high-pass filter* because it will pass high-frequency signals with little attenuation and block low-frequency signals.

PROJECT 6.1: The High-Pass Filter

OBJECTIVE

The objective of this project is to determine how V_{out} changes as the frequency of the input signal changes for a high-pass filter.

GENERAL INSTRUCTIONS

When the circuit is set up, measure V_{out} for each frequency; you will also calculate X_C for each frequency value to show the relationship between the output voltage and the reactance of the capacitor.

Parts List

You need the following equipment and supplies:

❑ One 1 kΩ, 0.25-watt resistor.

❑ One 0.016 µF capacitor. (You'll probably find 0.016 µF capacitors listed as polypropylene film capacitors. A *polypropylene film capacitor* is made with different material than the more typical ceramic capacitor but performs the same function. If your supplier doesn't carry 0.016 µF capacitors, you can use the closest value the supplier carries. Your results will be changed slightly but will show the same effect.)

❑ One function generator.

❑ One oscilloscope. (You can substitute a multimeter and measure V_{out} in rms voltage rather than peak-to-peak voltage.)

❑ One breadboard.

STEP-BY-STEP INSTRUCTIONS

Set up the circuit shown in Figure 6.10. If you have some experience in building circuits, this schematic (along with the previous parts list) should provide all the information you need to build the circuit. If you need a bit more help building the circuit, look at the photos of the completed circuit in the "Expected Results" section.

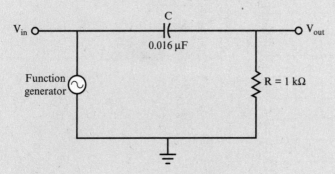

FIGURE 6.10

Carefully check your circuit against the diagram.

After you have checked your circuit, follow these steps, and record your measurements in the blank table following the steps:

1. Connect the oscilloscope probe for channel 2 to a jumper wire connected to V_{in}, and connect the ground clip to a jumper wire attached to the ground bus.

2. Connect the oscilloscope probe for channel 1 to a jumper wire connected to V_{out}, and connect the ground clip to a jumper wire attached to the ground bus.

3. Set the function generator to generate a 10 V$_{pp}$, 25 Hz sine wave.

4. Measure and record V$_{out}$.

5. Adjust the function generator to the frequency shown in the next row of the table.

6. Measure and record V$_{out}$.

7. Repeat steps 5 and 6 until you have recorded V$_{out}$ for the last row of the table.

8. Calculate the values of X$_C$ for each row and enter them in the table.

f$_{in}$	X$_C$	V$_{out}$
25 Hz		
50 Hz		
100 Hz		
250 Hz		
500 Hz		
1 kHz		
3 kHz		
5 kHz		
7 kHz		
10 kHz		
20 kHz		
30 kHz		
50 kHz		
100 kHz		
200 kHz		
500 kHz		
1 MHz		

9. In the blank graph shown in Figure 6.11, plot V$_{out}$ versus f$_{in}$ with the voltage on the vertical axis and the frequency on the X axis. The curve should have the same shape as the curve shown in Figure 6.9, but don't worry if your curve is shifted slightly to the right or left.

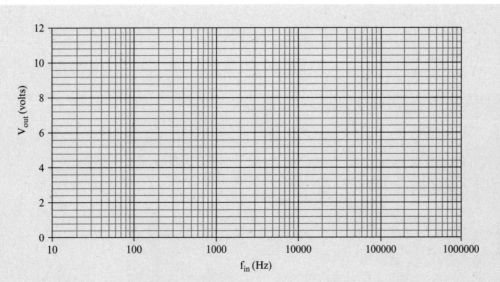

FIGURE 6.11

EXPECTED RESULTS

Figure 6.12 shows the breadboarded circuit for this project.

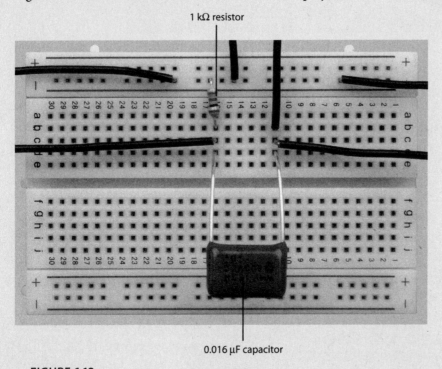

1 kΩ resistor

0.016 µF capacitor

FIGURE 6.12

Figure 6.13 shows a function generator and oscilloscope attached to the circuit.

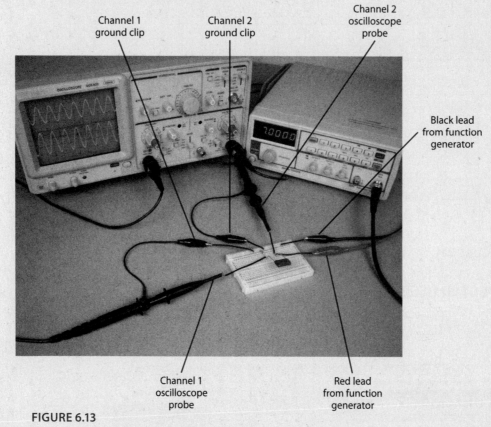

Channel 1
ground clip

Channel 2
ground clip

Channel 2
oscilloscope
probe

Black lead
from function
generator

Channel 1
oscilloscope
probe

Red lead
from function
generator

FIGURE 6.13

The input signal is represented by the upper sine wave shown in Figure 6.14, and the output signal is represented by the lower sine wave.

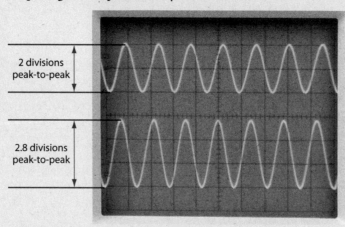

2 divisions
peak-to-peak

2.8 divisions
peak-to-peak

FIGURE 6.14

As you change f_{in}, you may need to adjust the TIME/DIV, VOLTS/DIV, and vertical POSITION controls. The controls shown in Figure 6.15 are adjusted to measure V_{out} when $f_{in} = 7$ kHz.

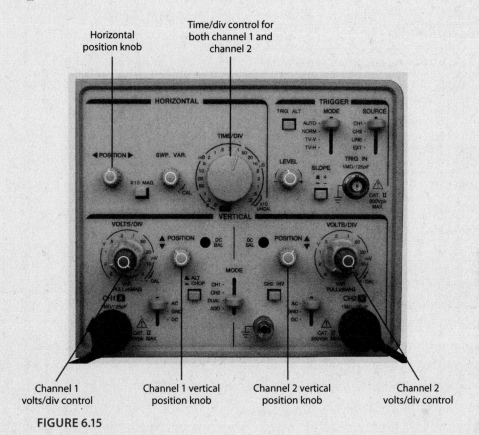

FIGURE 6.15

Your values should be close to those shown in the following table, and the curve should be similar to Figure 6.16:

f_{in}	X_C	V_{out}
25 Hz	400 kΩ	0.025 volts
50 Hz	200 kΩ	0.05 volts
100 Hz	100 kΩ	0.1 volts
250 Hz	40 kΩ	0.25 volts
500 Hz	20 kΩ	0.5 volts

Continued

(continued)

f$_{in}$	X$_C$	V$_{out}$
1 kHz	10 kΩ	1 volts
3 kHz	3.3 kΩ	2.9 volts
5 kHz	2 kΩ	4.5 volts
7 kHz	1.4 kΩ	5.6 volts
10 kHz	1 kΩ	7.1 volts
20 kHz	500 Ω	8.9 volts
30 kHz	330 Ω	9.5 volts
50 kHz	200 Ω	9.8 volts
100 kHz	100 Ω	10 volts
200 kHz	50 Ω	10 volts
500 kHz	20 Ω	10 volts
1 MHz	10 Ω	10 volts

Notice the relationship between X$_C$ and V$_{out}$ in this circuit. Low values of V$_{out}$ and the voltage drop across the resistor in this circuit occur at frequencies for which X$_C$ is high. When X$_C$ is high, more voltage is dropped across the capacitor, and less voltage is dropped across the resistor. (Remember that X$_C$ changes with frequency, while the value of the resistor stays constant.) Similarly, when X$_C$ is low, less voltage is dropped across the capacitor, and more voltage is dropped across the resistor, resulting in a higher V$_{out}$.

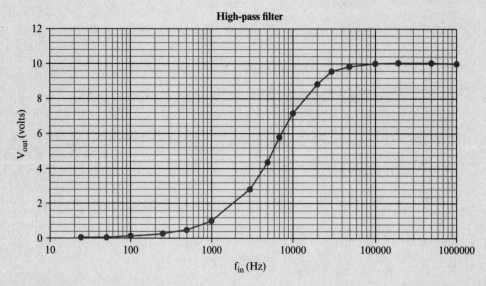

FIGURE 6.16

14 Refer to the curve you drew in Project 6.1 for the following question.

QUESTION

What would cause your curve to be moved slightly to the right or the left of the curve shown in Figure 6.16? _____

_____

ANSWER

Slightly different values for the resistor and capacitor that you used versus the resistor and capacitor used to produce the curve in Figure 6.16. Variations in resistor and capacitor values are to be expected, given the tolerance allowed for standard components.

15 The circuit shown in Figure 6.17 is used in many electronic devices.

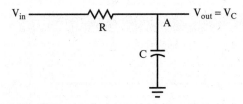

FIGURE 6.17

For this circuit, you measure the output voltage across the capacitor instead of across the resistor (between point A and ground).

The impedance of this circuit is the same as that of the circuit used in the last few problems. It still behaves like a voltage divider, and you can calculate the output voltage with an equation similar to the one you used for the high-pass filter circuit discussed in the last few problems. However, by switching the positions of the resistor and capacitor to create the circuit shown in Figure 6.17, you switch which frequencies will be attenuated, and which will not be attenuated, making the new circuit a low-pass filter, whose characteristics you explore in the next few problems.

QUESTIONS

A. What is the impedance formula for the circuit? _____

_____

B. What is the formula for the output voltage? _____

_____

ANSWERS

A. $Z = \sqrt{X_C^2 + R^2}$

B. $V_{out} = V_{in} \times \dfrac{X_C}{Z}$

16 Refer to the circuit shown in Figure 6.17 and the following values:

$V_{in} = 10\,V_{pp}, f = 2\,kHz$

$C = 0.1\mu F, R = 1k\Omega$

QUESTIONS

Find the following:

A. X_C: _____

_____

B. Z: _____

_____

C. V_{out}: _____

_____

ANSWERS

A. 795 ohms

B. 1277 ohms

C. 6.24 volts

17 Again, refer to the circuit shown in Figure 6.17 to answer the following question.

QUESTION

Calculate the voltage across the resistor using the values given in problem 16, along with the calculated impedance value. _____

_____

ANSWER

$$V_R = V_{in} \times \frac{R}{Z} = 10 \times \frac{1000}{1277} = 7.83V_{pp}$$

18 Use the information from problems 16 and 17 to answer the following question.

QUESTION

What is the formula to calculate V_{in} using the voltages across the capacitor and the resistor? _____

_____

ANSWER

The formula is $V_{in}^2 = V_C^2 + V_R^2$.

19 V_{out} of the circuit shown in Figure 6.17 changes as the frequency of the input signal changes. Figure 6.18 shows the graph of V_{out} versus frequency for this circuit.

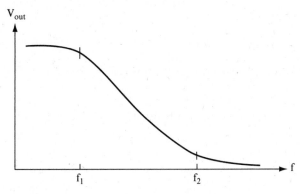

FIGURE 6.18

QUESTION

What parameters determine f_1 and f_2? _____

_____

ANSWER

The values of the capacitor and the resistor

NOTE You can see in Figure 6.18 that V_{out} is large for the lowest frequency and nearly zero for the highest frequency. This type of circuit is called a *low-pass filter* because it will pass low frequency signals with little attenuation, while blocking high-frequency signals.

PROJECT 6.2: The Low-Pass Filter

OBJECTIVE

The objective of this project is to determine how V_{out} changes as the frequency of the input signal changes for a low-pass filter.

GENERAL INSTRUCTIONS

After the circuit is set up, measure V_{out} for each frequency. You also calculate X_C for each frequency value to show the relationship between the output voltage and the reactance of the capacitor.

Parts List

You need the following equipment and supplies:

❑ One 1 kΩ, 0.25-watt resistor. (You can use the same resistor that you used in Project 6.1.)

❑ One 0.016 µF capacitor. (You can use the same capacitor that you used in Project 6.1.)

- ❏ One function generator.
- ❏ One oscilloscope. (You can substitute a multimeter and measure V_{out} in rms voltage rather than peak-to-peak voltage.)
- ❏ One breadboard.

STEP-BY-STEP INSTRUCTIONS

Set up the circuit shown in Figure 6.19. If you have some experience in building circuits, this schematic (along with the previous parts list) should provide all the information you need to build the circuit. If you need a bit more help building the circuit, look at the photos of the completed circuit in the "Expected Results" section.

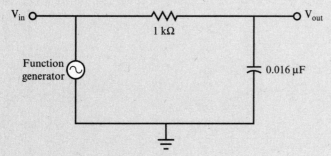

FIGURE 6.19

Carefully check your circuit against the diagram.

After you have checked your circuit, follow these steps, and record your measurements in the blank table following the steps:

1. Connect the oscilloscope probe for channel 2 to a jumper wire connected to V_{in}, and connect the ground clip to a jumper wire attached to the ground bus.

2. Connect the oscilloscope probe for channel 1 to a jumper wire connected to V_{out}, and connect the ground clip to a jumper wire attached to the ground bus.

3. Set the function generator to generate a 10 V_{pp}, 25 Hz sine wave.

4. Measure and record V_{out}.

5. Adjust the function generator to the frequency shown in the next row of the table.

6. Measure and record V_{out}.

7. Repeat steps 5 and 6 until you have recorded V_{out} for the last row of the table.

8. Enter the values of X_C for each row in the table. (Because you used the same capacitor and resistor in Project 6.1, you can take the values X_C from the table in Project 6.1.)

f_{in}	X_C	V_{out}
25 Hz		
50 Hz		
100 Hz		
250 Hz		
500 Hz		
1 kHz		
3 kHz		
5 kHz		
7 kHz		
10 kHz		
20 kHz		
30 kHz		
50 kHz		
100 kHz		
200 kHz		
500 kHz		
1 MHz		

9. In the blank graph shown in Figure 6.20, plot V_{out} versus f_{in} with the voltage on the vertical axis and the frequency on the X axis. The curve should have the same shape as the curve shown in Figure 6.18.

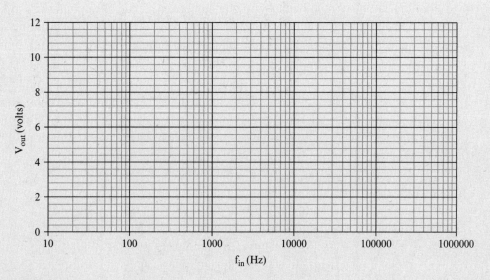

FIGURE 6.20

EXPECTED RESULTS

Figure 6.21 shows the breadboarded circuit for this project.

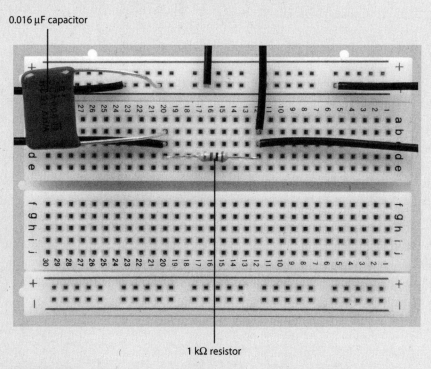

FIGURE 6.21

Figure 6.22 shows a function generator and oscilloscope attached to the circuit.

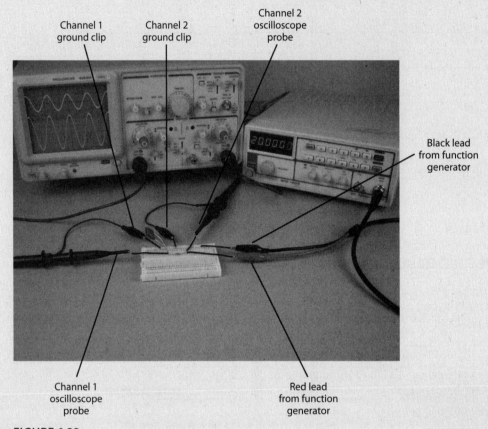

Channel 1
ground clip

Channel 2
ground clip

Channel 2
oscilloscope
probe

Black lead
from function
generator

Channel 1
oscilloscope
probe

Red lead
from function
generator

FIGURE 6.22

The input signal is represented by the upper sine wave, as shown in Figure 6.23, and the output signal is represented by the lower sine wave. Reading the number of divisions for the peak-to-peak output sine wave and multiplying it by the corresponding VOLTS/DIV setting allows to you measure V_{out}.

As you change f_{in} adjustments in the TIME/DIV control, the VOLTS/DIV and vertical POSITION controls for channel 1 may be needed. The controls shown in Figure 6.24 are adjusted to measure V_{out} when $f_{in} = 20$ kHz.

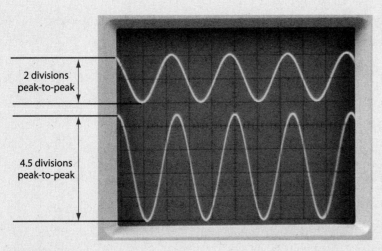

2 divisions
peak-to-peak

4.5 divisions
peak-to-peak

FIGURE 6.23

Horizontal
position knob

Time/div control for
both channel 1 and
channel 2

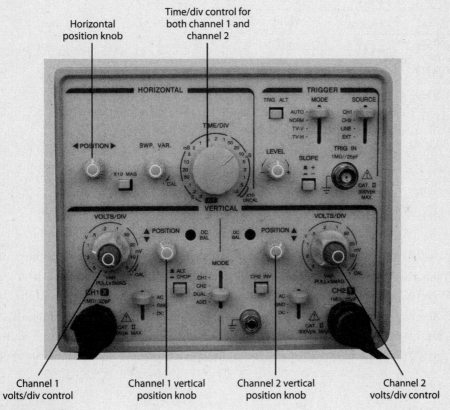

Channel 1
volts/div control

Channel 1 vertical
position knob

Channel 2 vertical
position knob

Channel 2
volts/div control

FIGURE 6.24

Your values should be close to those shown in the following table, and the curve should be similar to Figure 6.25.

f_{in}	X_C	V_{out}
25 Hz	400 kΩ	10 volts
50 Hz	200 kΩ	10 volts
100 Hz	100 kΩ	10 volts
250 Hz	40 kΩ	10 volts
500 Hz	20 kΩ	10 volts
1 kHz	10 kΩ	10 volts
3 kHz	3.3 kΩ	9.4 volts
5 kHz	2 kΩ	9.1 volts
7 kHz	1.4 kΩ	8.2 volts
10 kHz	1 kΩ	7.1 volts
20 kHz	500 Ω	4.5 volts
30 kHz	330 Ω	2.9 volts
50 kHz	200 Ω	2.0 volts
100 kHz	100 Ω	1 volt
200 kHz	50 Ω	0.5 volt
500 kHz	20 Ω	0.2 volt
1 MHz	10 Ω	0.1 volt

Notice the relationship between X_C and V_{out} in this circuit. Low values of V_{out} (The voltage drop across the capacitor in this circuit.) occur at frequencies for which X_C is also low. When X_C is low, more voltage is dropped across the resistor and less across the capacitor. (Remember that X_C changes with frequency, whereas the value of the resistor stays constant.) Similarly, when X_C is high, less voltage is dropped across the resistor, and more voltage is dropped across the capacitor, resulting in a higher V_{out}.

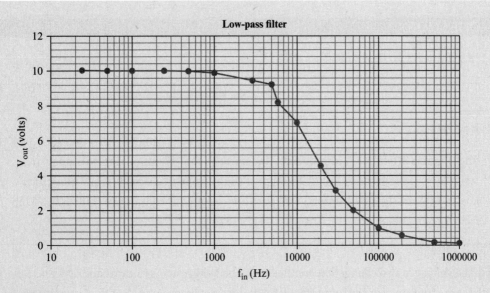

FIGURE 6.25

PHASE SHIFT OF AN RC CIRCUIT

20 In both of the circuits shown in Figure 6.26, the output voltage is different from the input voltage.

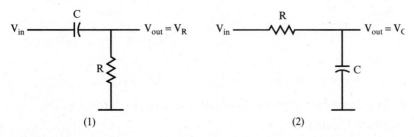

FIGURE 6.26

QUESTION

In what ways do they differ? _____

_____

ANSWER

The signal is attenuated, or reduced. The amount of attenuation depends upon the frequency of the signal. Circuit 1 will pass high-frequency signals while blocking low-frequency signals. Circuit 2 will pass low-frequency signals while blocking high-frequency signals.

21 The voltage is also changed in another way. The voltage across a capacitor rises and falls at the same frequency as the input signal, but it does not reach its peak at the same time, nor does it pass through zero at the same time. You can see this when you compare the V_{out} curves to the V_{in} curves in Figure 6.27.

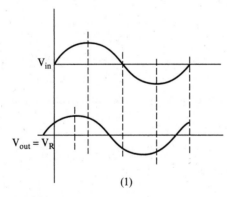
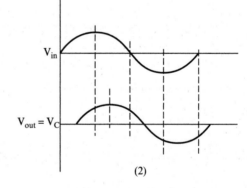

(1)   (2)

FIGURE 6.27

> **NOTE**   The numbered graphs in Figure 6.27 are produced by the corresponding numbered circuits in Figure 6.26.

QUESTIONS

A. Examine graph (1). Is the output voltage peak displaced to the right or the left?

_____

B. Examine graph (2). Is the output voltage peak displaced to the right or the left?

_____

22 The output voltage waveform in graph (1) of Figure 6.27 is said to *lead the input voltage waveform*. The output waveform in graph (2) is said to *lag the input waveform*. The amount that V_{out} leads or lags V_{in} is measured in degrees. There are 90 degrees between the peak of a sine wave and a point at which the sine wave crosses zero volts. You can use this information to estimate the number of degrees V_{out} is leading or lagging V_{in}. The difference between these two waveforms is called a *phase shift* or *phase difference*.

QUESTIONS

A. What is the approximate phase shift of the two waveforms shown in the graphs? __

_____

B. Do you think that the phase shift depends on the value of frequency? _____

_____

C. Will an RC voltage divider with the voltage taken across the capacitor produce a lead or a lag in the phase shift of the output voltage? _____

_____

ANSWERS

A. Approximately 35 degrees.

B. It does depend upon frequency because the values of the reactance and impedance depend upon frequency.

C. A lag as shown in graph (2).

23 The current through a capacitor is out of phase with the voltage across the capacitor. The current leads the voltage by 90 degrees. The current and voltage across a resistor are in phase. (That is, they have no phase difference.)

Figure 6.28 shows the vector diagram for a series RC circuit. θ is the phase angle by which V_R leads V_{in}. φ is the phase angle by which V_C lags V_{in}.

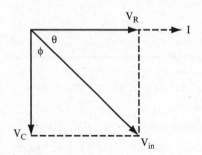

FIGURE 6.28

> **NOTE** Although the voltage across a resistor is in phase with the current through the resistor, both are out of phase with the applied voltage.

You can calculate the phase angle by using this formula:

$$\tan\theta = \frac{V_C}{V_R} = \frac{1}{2\pi fRC} = \frac{X_C}{R}$$

As an example, calculate the phase angle when 160 Hz is applied to a 3.9 kΩ resistor in series with a 0.1 μF capacitor.

$$\tan\theta = \frac{1}{2 \times \pi \times 160 \times 3.9 \times 10^3 \times 0.1 \times 10^{-6}} = 2.564$$

You can calculate the inverse tangent of 2.564 on your calculator and find that the phase angle is 68.7 degrees, which means that V_R leads V_{in} by 68.7 degrees. This also means that V_C lags the input by 21.3 degrees.

In electronics, the diagram shown in Figure 6.28 is called a *phasor diagram*, but the mathematics involved are the same as for vector diagrams, with which you should be familiar.

QUESTION

Sketch a phasor diagram using the angles θ and φ resulting from the calculations in this problem. Use a separate sheet of paper for your diagram.

ANSWER

See Figure 6.29. Note that the phasor diagram shows that the magnitude of V_C is greater than V_R.

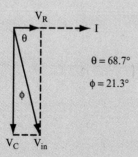

$$\theta = 68.7°$$

$$\phi = 21.3°$$

FIGURE 6.29

24 Using the component values and input signal shown in Figure 6.30, answer the following questions.

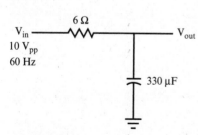

FIGURE 6.30

QUESTIONS

Find the following:

A. X_C: _____

_____

B. Z: _____

_____

C. V_{out}: _____

_____

D. V_R: _____

_____

E. The current flowing through the circuit: _____

_____

F. The phase angle: _____

_____

ANSWERS

A. $X_C = \dfrac{1}{2\pi fC} = 8\,\text{ohms}$

B. $Z = \sqrt{8^2 + 6^2} = 10\,\text{ohms}$

C. $V_{out} = V_C = V_{in} \times \dfrac{X_C}{Z} = 8\,\text{volts}$

D. $V_R = V_{in} \times \dfrac{R}{Z} = 6\,\text{volts}$

E. $I = \dfrac{V}{Z} = \dfrac{10\,V_{pp}}{10\,\Omega} = 1\,\text{ampere}$

F. $\tan\theta = \dfrac{X_C}{R} = \dfrac{8\,\Omega}{6\,\Omega} = 1.33.$

Therefore, $\theta = 53.13$ degrees.

25 Use the circuit shown in Figure 6.31 to answer the following questions.

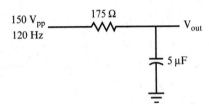

FIGURE 6.31

QUESTIONS

Calculate the following parameters:

A. X_C: _____

_____

B. Z: _____

_____

C. V_{out}: _____

_____

D. V_R: _____

_____

E. The current flowing through the circuit: _____

_____

F. The phase angle: _____

_____

ANSWERS

A. X_C = 265 ohms

B. $Z = \sqrt{175^2 + 265^2} = 317.57\,\Omega$

C. V_C = 125 volts

D. V_R = 83 volts

E. I = 0.472 ampere

F. $\tan\theta = \dfrac{265\,\Omega}{175\,\Omega} = 1.5$

Therefore, θ = 56.56 degrees.

RESISTOR AND CAPACITOR IN PARALLEL

26 The circuit shown in Figure 6.32 is a common variation on the low-pass filter circuit introduced in problem 15.

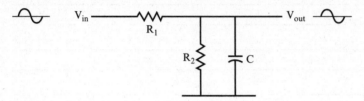

FIGURE 6.32

Because a DC signal will not pass through the capacitor, this circuit functions like the circuit shown in Figure 6.33 for DC input signals.

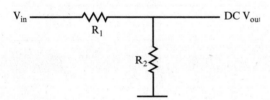

FIGURE 6.33

An AC signal will pass through both the capacitor and R_2. You can treat the circuit as if it had a resistor with a value of r (where r is the parallel equivalent of R_2 and X_c) in place of the parallel capacitor and resistor. This is shown in Figure 6.34.

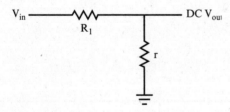

FIGURE 6.34

Calculating the exact parallel equivalent (r) is complicated and beyond the scope of this book. However, to demonstrate the usefulness of this circuit, you can make a major simplification. Consider a circuit where X_C is only about one-tenth the value of R_2 or less. This circuit has many practical applications, because it attenuates the AC and the DC differently.

The following example can help to clarify this. For the following circuit, calculate the AC and DC output voltages separately.

For the circuit shown in Figure 6.35, you can calculate the AC and DC output voltages separately by following the steps outlined in the following questions.

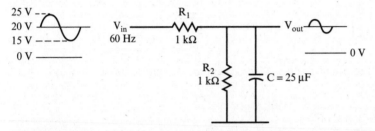

FIGURE 6.35

QUESTIONS

A. Find X_C. Check that it is less than one-tenth of R_2.

$$X_C = \underline{\hspace{8cm}}$$

B. For the circuit in Figure 6.35, determine through which circuit components DC signals will flow. Then use the voltage divider formula to find DC V_{out}.

DC $V_{out} = \underline{\hspace{6cm}}$

C. For the circuit in Figure 6.35 determine which circuit components AC signals will flow through. Then use the voltage divider formula to find AC V_{out}.

AC $V_{out} = \underline{\hspace{6cm}}$

D. Compare the AC and DC input and output voltages. $\underline{\hspace{5cm}}$

$\underline{\hspace{11cm}}$

ANSWERS

A. $X_C = 106$ ohms and $R_2 = 1000$ ohms, so X_C is close enough to one-tenth of R_2.

B. Figure 6.36 shows the portion of the circuit that a DC signal passes through.

$$V_{out} = 20 \times \frac{1k\Omega}{1k\Omega + 1k\Omega} = 10 \text{ volts}$$

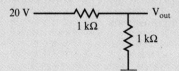

FIGURE 6.36

C. Figure 6.37 shows the portion of the circuit that an AC signal passes through.

$$V_{out} = 10 \times \frac{106}{\sqrt{(1000)^2 + (106)^2}} = 1.05 \text{ volts}$$

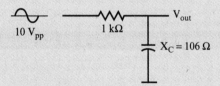

FIGURE 6.37

D. Figure 6.38 shows the input waveform on the left and the output waveform on the right. You can see from the waveforms that the DC voltage has dropped from 20 volts to 10 volts and that the AC voltage has dropped from 10 volts to 1.05 volts.

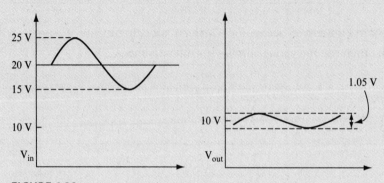

FIGURE 6.38

27 Figure 6.39 shows two versions of the circuit discussed in problem 26 with changes to the value of the capacitor or the frequency of the input signal. The DC input voltage is 20 volts, and the AC input voltage is 10 V_{pp}. Use the same steps shown in problem 26 to find and compare the output voltages with the input voltages for the two circuits shown in Figure 6.39.

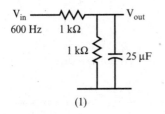
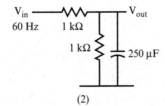

FIGURE 6.39

QUESTIONS

1.

   A. $X_C =$ _____

   B. DC $V_{out} =$ _____

   C. AC $V_{out} =$ _____

   D. Attenuation: _____

2.

   A. $X_C =$ _____

   B. DC $V_{out} =$ _____

   C. AC $V_{out} =$ _____

   D. Attenuation: _____

ANSWERS

1.

   A. X_C = 10.6 ohms.

   B. DC V_{out} = 10 volts.

   C. AC V_{out} = 0.1 volts.

   D. Here, the DC attenuation is the same as the example in problem 26, but the AC output voltage is reduced because of the higher frequency.

Continued

(continued)

ANSWERS

2.

    A. X_C = 10.6 ohms.

    B. DC V_{out} = 10 volts.

    C. AC V_{out} = 0.1 volts.

    D. The DC attenuation is still the same, but the AC output voltage is reduced because of the larger capacitor. *(compared to prob. 26)*

INDUCTORS IN AC CIRCUITS

28 Figure 6.40 shows a voltage divider circuit using an inductor, rather than a capacitor.

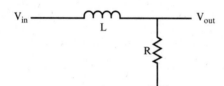

FIGURE 6.40

As with previous problems, consider all the inputs to be pure sine waves. Like the capacitor, the inductor cannot change the frequency of a sine wave, but it can reduce the amplitude of the output voltage.

The simple circuit, as shown in Figure 6.40, opposes current flow.

QUESTIONS

    A. What is the opposition to current flow called? _____

    _____

    B. What is the formula for the reactance of the inductor? _____

    _____

    C. Write out the formula for the opposition to the current flow for this circuit. _____

    _____

ANSWERS

A. Impedance

B. $X_L = 2\pi fL$.

C. $Z = \sqrt{X_L^2 + R^2}$

In many cases, the DC resistance of the inductor is low, so assume that it is 0 ohms. For the next two problems, make that assumption in performing your calculations.

29 You can calculate the voltage output for the circuit shown in Figure 6.41 with the voltage divider formula.

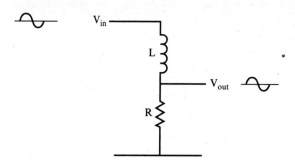

FIGURE 6.41

QUESTION

What is the formula for V_{out}? _____

_____

ANSWER

$$V_{out} = V_{in} \times \frac{R}{Z}$$

30 Find the output voltage for the circuit shown in Figure 6.42.

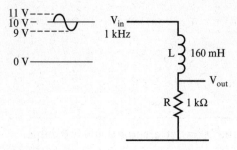

FIGURE 6.42

Use the steps in the following questions to perform the calculation.

QUESTIONS

A. Find the DC output voltage. Use the DC voltage divider formula.

DC V_{out} = ——————————————————

B. Find the reactance of the inductor.

X_L = ——————————————————

C. Find the AC impedance.

Z = ——————————————————

D. Find the AC output voltage.

AC V_{out} = ——————————————————

E. Combine the outputs to find the actual output. Draw the output waveform and label the voltage levels of the waveform on the blank graph shown in Figure 6.43.

FIGURE 6.43

ANSWERS

A. DC $V_{out} = 10$ volts $\times \dfrac{1k\Omega}{1k\Omega + 0} = 10$ volts

B. $X_L = 1\ k\Omega$ (approximately).

C. $Z = \sqrt{1^2 + 1^2} = \sqrt{2} = 1.414\,k\Omega$

D. AC $V_{out} = 2V_{pp} \times \dfrac{1k\Omega}{1.414\,k\Omega} = 1.414\,V_{pp}$

E. The output waveform is shown in Figure 6.44.

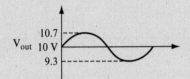

FIGURE 6.44

31 For the circuit shown in Figure 6.45, the DC resistance of the inductor is large enough that you should include that value in your calculations.

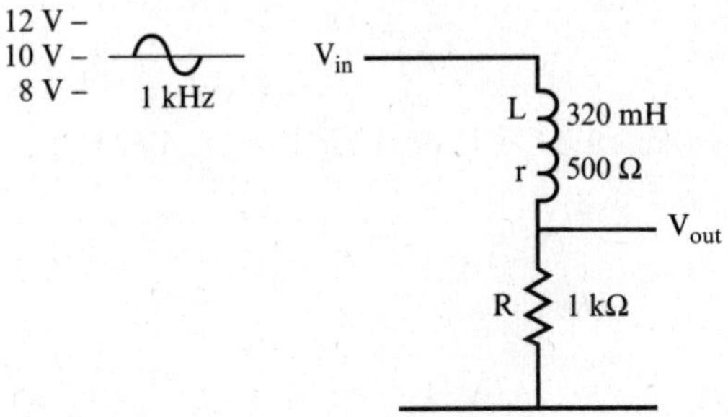

12 V –
10 V – V_{in}
8 V –
1 kHz

L 320 mH
r 500 Ω
V_{out}
R 1 kΩ

FIGURE 6.45

QUESTIONS

For the circuit shown in Figure 6.45, calculate the DC and AC output voltages, using the steps listed in problem 30.

A. DC $V_{out} = $ _____

B. $X_L = $ _____

C. Z = _____

D. AC V$_{out}$ = _____

E. Draw the output waveform and label the voltage levels of the waveform on the blank graph in Figure 6.46.

FIGURE 6.46

ANSWERS

A. DC V$_{out}$ = 10 volts = $\dfrac{1\,k\Omega}{(1\,k\Omega + 500\,\Omega)}$ = 6.67 volts

NOTE The 500 Ω DC resistance of the inductor has been added to the 1 kΩ resistor value in this calculation.

B. X$_{L}$ = 2 kΩ,

C. Z = $\sqrt{1.5^2 + 2^2}$ = 2.5 kΩ

NOTE The 500 Ω DC resistance of the inductor has been added to the 1 kΩ resistor value in this calculation.

D. AC V$_{out}$ = 1.6 V$_{pp}$,

E. See Figure 6.47.

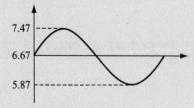

FIGURE 6.47

32 To calculate V_{out}, in problems 30 and 31, you also had to calculate X_L. However, because X_L changes with the frequency of the input signal, the impedance and the amplitude of V_{out} also change with the frequency of the input signal. If you plot the output voltage V_{out} against frequency, you should see the curve shown in Figure 6.48.

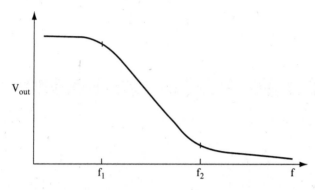

FIGURE 6.48

The values of the inductor and resistor determine the frequency at which V_{out} starts to drop (f_1), and the frequency at which V_{out} levels off (f_2).

The curve in Figure 6.48 shows that using an inductor and resistor in a circuit such as the one shown in Figure 6.42 produces a low-pass filter similar to the one discussed in problems 15 through 19.

QUESTION

What values control f_1 and f_2? _____

_____

ANSWER

The values of the inductor and the resistor

33 You can also create a circuit as shown in Figure 6.49, in which the output voltage is equal to the voltage drop across the inductor.

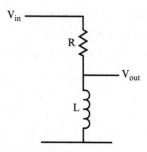

FIGURE 6.49

QUESTIONS

A. What formula would you use to find V_{out}? _____

_____

B. If you plot the output voltage versus the frequency, what would you expect the curve to be? Use a separate sheet of paper to draw your answer. _____

_____

ANSWERS

A. $V_{out} = V_{in} \times \dfrac{X_L}{Z}$

B. See Figure 6.50.

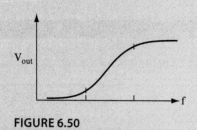

FIGURE 6.50

The curve in Figure 6.50 demonstrates that using an inductor and resistor in a circuit, such as the one shown in Figure 6.49, produces a high-pass filter similar to the one discussed in problems 6 through 13.

HIGHER-ORDER FILTERS

Filter circuits that contain one capacitor or inductor are called *first-order filters*. Filter order numbers reflect the number of capacitors, inductors, or operational amplifiers (a component discussed in Chapter 8, "Transistor Amplifiers") in the filter. For example, a filter that contains four capacitors is a fourth-order filter, whereas a filter that contains six capacitors is a sixth-order filter.

If you want a sharper drop-off between frequencies, you can connect first-order filters in series. This effect is demonstrated in the following figure.

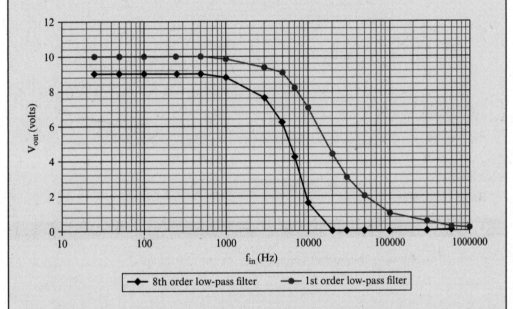

This graph shows how V_{out} changes as f_{in} changes for the first-order low-pass filter used in Project 6.2, and for an eighth-order low-pass filter. V_{out} for the eighth-order filter drops by 80 percent at approximately 10 kHz, whereas V_{out} for the first-order filter doesn't drop by 80 percent until the frequency reaches approximately 50 kHz.

PHASE SHIFT FOR AN RL CIRCUIT

34 Filter circuits that use inductors (such as those shown in Figure 6.51) produce a phase shift in the output signal, just as filter circuits containing capacitors do. You can see the shifts for the circuits shown in Figure 6.51 by comparing the input and output waveforms shown below the circuit diagrams.

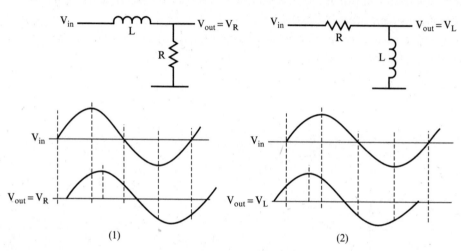

FIGURE 6.51

QUESTION

In which circuit does the output voltage lead the input voltage? _____

_____

ANSWER

In graph (1), the output voltage lags the input voltage, and in graph (2), the output voltage leads.

35 Figure 6.52 shows a vector diagram for both the circuits shown in Figure 6.51. The current through the inductor lags the voltage across the inductor by 90 degrees.

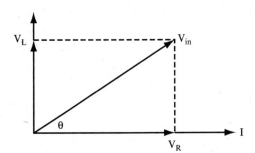

FIGURE 6.52

The phase angle is easily found:

$$\tan\theta = \frac{V_L}{V_R} = \frac{X_L}{R} = \frac{2\pi fL}{R}$$

QUESTION

Calculate the phase angle for the circuit discussed in problem 30. _____

_____

ANSWER

45 degrees

36 Refer to the circuit discussed in problem 31.

QUESTION

Calculate the phase angle. _____

_____

ANSWER

$$\tan\theta = \frac{X_L}{R} = \frac{2\,k\Omega}{1.5\,k\Omega} = 1.33$$

Therefore, $\theta = 53.1$ degrees.

SUMMARY

This chapter has discussed the uses of capacitors, resistors, and inductors in voltage divider and filter circuits. You learned how to determine the following:

- The output voltage of an AC signal after it passes through a high-pass RC filter circuit

- The output voltage of an AC signal after it passes through a low-pass RC circuit

- The output voltage of an AC signal after it passes through a high-pass RL circuit

- The output voltage of an AC signal after it passes through a low-pass RL circuit

- The output waveform of an AC or combined AC-DC signal after it passes through a filter circuit

- Simple phase angles and phase differences

SELF-TEST

These questions test your understanding of this chapter. Use a separate sheet of paper for your calculations. Compare your answers with the answers provided following the test.

For questions 1–3, calculate the following parameters for the circuit shown in each question.

A. X_C

B. Z

C. V_{out}

D. I

E. $\tan \theta$ and θ

1. Use the circuit shown in Figure 6.53.

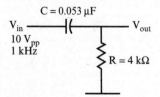

FIGURE 6.53

A. _____

B. _____

C. _____

D. _____

E. _____

2. Use the circuit shown in Figure 6.54.

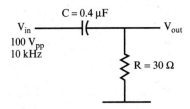

C = 0.4 μF

V_{in}
100 V_{pp}
10 kHz

V_{out}

R = 30 Ω

FIGURE 6.54

A. _____

B. _____

C. _____

D. _____

E. _____

3. Use the circuit shown in Figure 6.55.

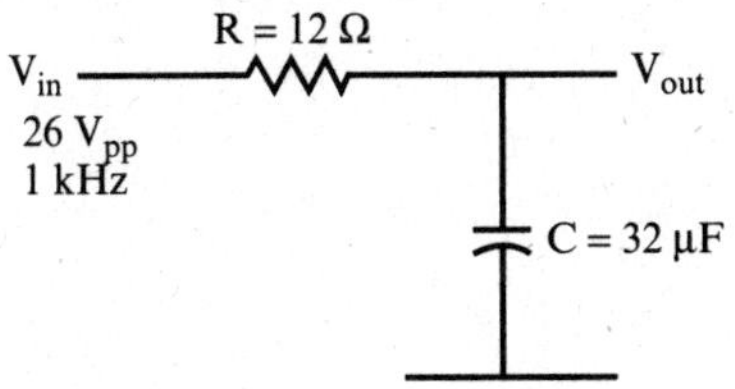

R = 12 Ω

V_{in}
26 V_{pp}
1 kHz

V_{out}

C = 32 μF

FIGURE 6.55

A. _____

B. _____

C. _____

D. _____

E. _____

For questions 4–6, calculate the following parameters for the circuit shown in each question.

    A. X_C

    B. AC V_{out}

    C. DC V_{out}

4. Use the circuit shown in Figure 6.56.

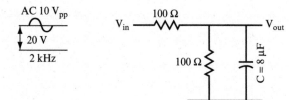

FIGURE 6.56

    A. _____

    B. _____

    C. _____

5. Use the circuit shown in Figure 6.57.

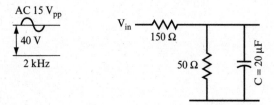

FIGURE 6.57

    A. _____

    B. _____

    C. _____

6. Use the circuit shown in Figure 6.58.

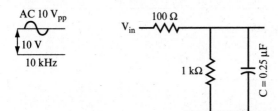

FIGURE 6.58

A. _____

B. _____

C. _____

For questions 7–9, calculate the following parameters for the circuit shown in each question.

    A. DC V_{out}

    B. X_L

    C. Z

    D. AC V_{out}

    E. $\tan \theta$ and θ

7. Use the circuit shown in Figure 6.59.

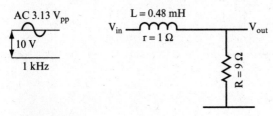

FIGURE 6.59

A. _____

B. _____

C. _____

D. _____

E. _____

8. Use the circuit shown in Figure 6.60.

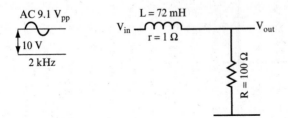

FIGURE 6.60

A. _____

B. _____

C. _____

D. _____

E. _____

9. Use the circuit shown in Figure 6.61.

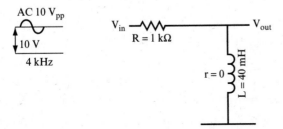

FIGURE 6.61

A. _____

B. _____

C. _____

D. _____

E. _____

ANSWERS TO SELF-TEST

If your answers do not agree with those provided here, review the applicable problems in this chapter before you go to Chapter 7.

1A.	3 kΩ	problems 8, 9, 10, 23
1B.	5 kΩ	
1C.	8 volts	
1D.	2 amperes	
1E.	36.87 degrees	
2A.	40 ohms	problems 8, 9, 23
2B.	50 ohms	
2C.	60 volts	
2D.	2 amperes	
2.E	53.13 degrees	
3A.	5 ohms	problems 8, 9, 23
3B.	13 ohms	
3C.	10 volts	
3D.	2 amperes	
3E.	22.63 degrees	
4A.	10 ohms	problems 26 and 27
4B.	1 volt	
4C.	10 volts	
5A.	4 ohms	problems 26 and 27
5B.	0.4 volt	
5C.	10 volts	
6A.	64 ohms	problems 26 and 27
6B.	5.4 volts	
6C.	9.1 volts	
7A.	9 volts	problems 28–30, 35
7B.	3 ohms	
7C.	10.4 ohms	
7D.	2.7 volts	

Continued

(continued)

7E.	16.7 degrees	
8A.	10 volts	problems 28–30, 35
8B.	904 ohms	
8C.	910 ohms	
8D.	1 volt	
8E.	83.69 degrees	
9A.	0 volts	problems 28–30, 35
9B.	1 kΩ	
9C.	1.414 kΩ	
9D.	7 volts	
9E.	45 degrees	

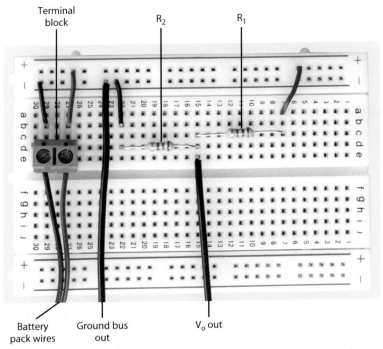

VOLTAGE DIVIDER CIRCUIT ASSEMBLED ON A BREADBOARD (CHAPTER 1)

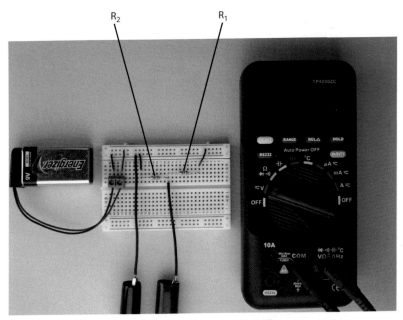

MULTIMETER CONNECTED TO A VOLTAGE DIVIDER CIRCUIT TO MEASURE VOLTAGE (CHAPTER 1)

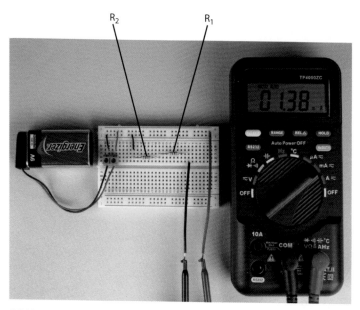

CONNECTING A MULTIMETER TO A VOLTAGE DIVIDER CIRCUIT TO MEASURE CURRENT
(CHAPTER 1)

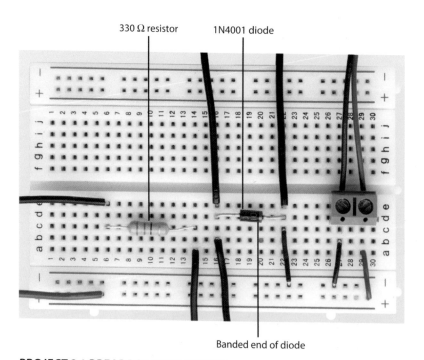

PROJECT 2.1 BREADBOARDED CIRCUIT

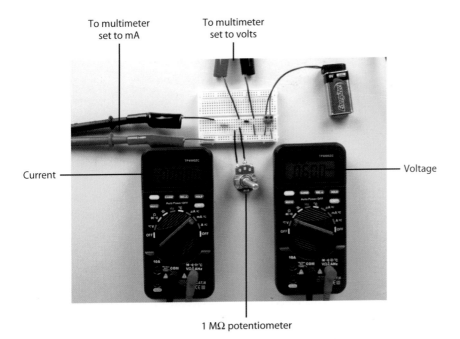

To multimeter set to mA

To multimeter set to volts

Current

Voltage

1 MΩ potentiometer

PROJECT 2.1 TEST SETUP

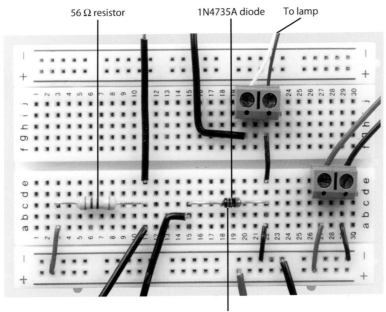

56 Ω resistor

1N4735A diode

To lamp

Banded end of diode

PROJECT 2.2 BREADBOARDED CIRCUIT

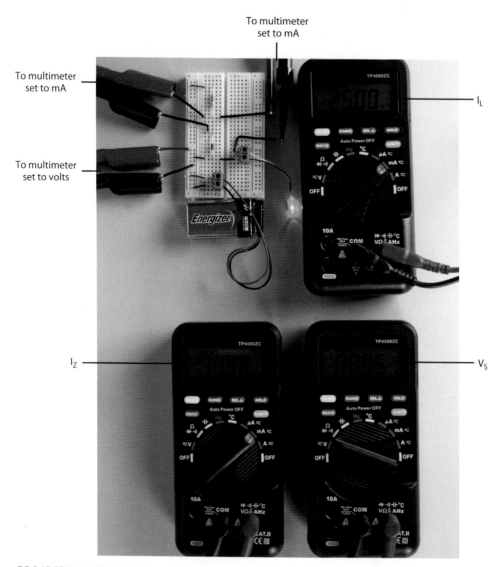

To multimeter set to mA

To multimeter set to mA

To multimeter set to volts

I_L

I_Z

V_S

PROJECT 2.2 TEST SETUP

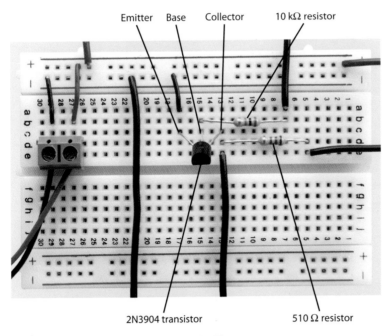

PROJECT 3.1 BREADBOARDED CIRCUIT

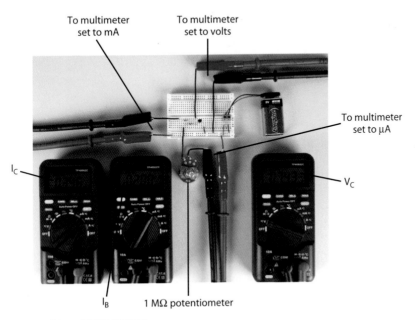

PROJECT 3.1 TEST SETUP

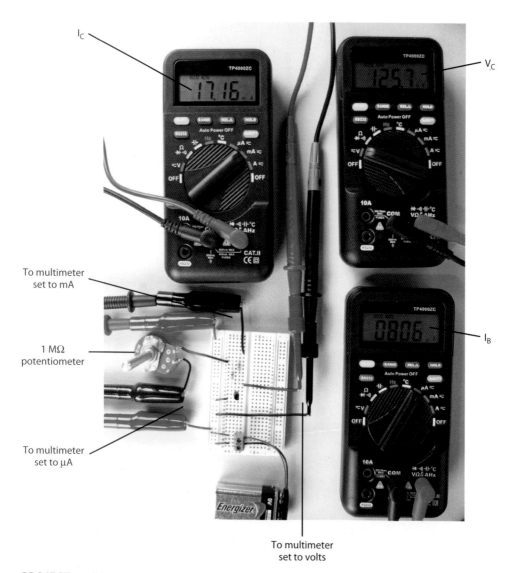

I_C

V_C

To multimeter
set to mA

1 MΩ
potentiometer

I_B

To multimeter
set to µA

To multimeter
set to volts

PROJECT 3.2 TEST SETUP

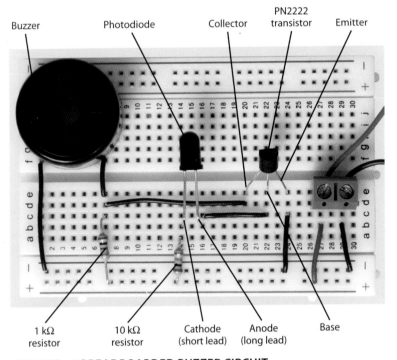

PROJECT 4.1 BREADBOARDED BUZZER CIRCUIT

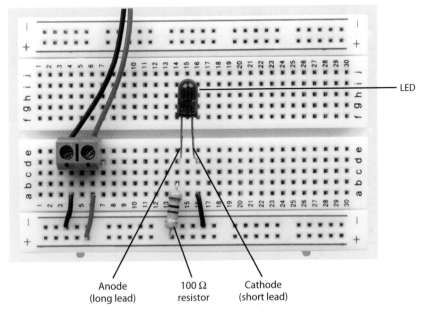

PROJECT 4.1 BREADBOARDED LED CIRCUIT

PROJECT 4.1 TEST SETUP

Drain   Gate   Source

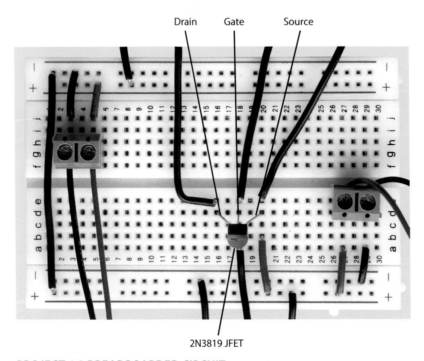

2N3819 JFET

PROJECT 4.2 BREADBOARDED CIRCUIT

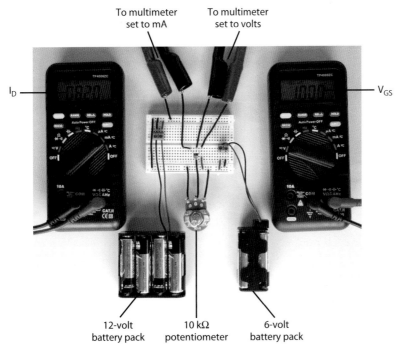

To multimeter set to mA

To multimeter set to volts

I_D

V_{GS}

12-volt battery pack

10 kΩ potentiometer

6-volt battery pack

PROJECT 4.2 TEST SETUP

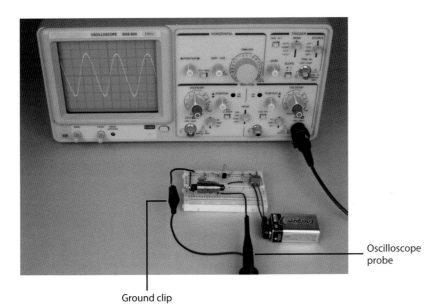

Oscilloscope probe

Ground clip

OSCILLOSCOPE WHOSE PROBE CONNECTS TO THE OUTPUT OF AN OSCILLATOR CIRCUIT (CHAPTER 5)

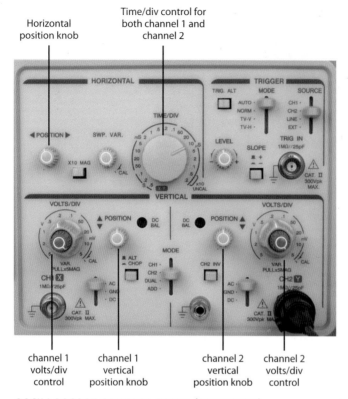

Horizontal
position knob

Time/div control for
both channel 1 and
channel 2

channel 1
volts/div
control

channel 1
vertical
position knob

channel 2
vertical
position knob

channel 2
volts/div
control

OSCILLOSCOPE CONTROL PANEL (CHAPTER 5)

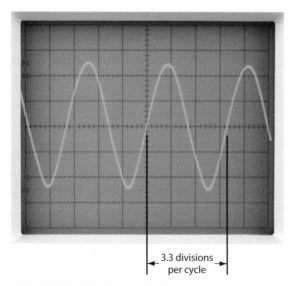

3.3 divisions
per cycle

THE PERIOD OF THE SINE WAVE GENERATED BY THE OSCILLATOR CIRCUIT (CHAPTER 5)

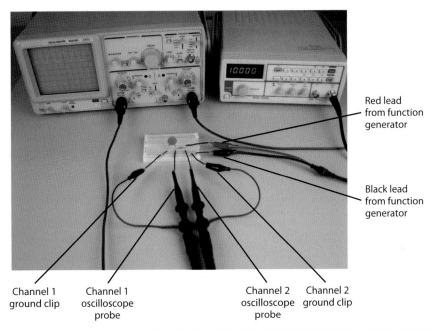

Red lead from function generator

Black lead from function generator

Channel 1 ground clip

Channel 1 oscilloscope probe

Channel 2 oscilloscope probe

Channel 2 ground clip

FUNCTION GENERATOR AND OSCILLOSCOPE ATTACHED TO THE VOLTAGE DIVIDER CIRCUIT (CHAPTER 5)

Phase shift

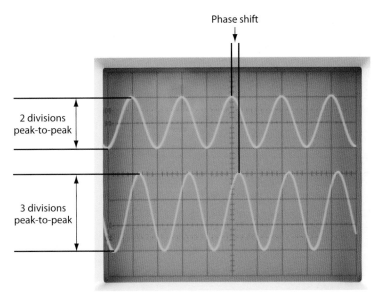

2 divisions peak-to-peak

3 divisions peak-to-peak

ADJUSTING THE VOLT/DIV CONTROLS AND VERTICAL POSITION CONTROLS FOR CHANNELS 1 AND 2 TO FIT BOTH SINE WAVES ON THE SCREEN (CHAPTER 5)

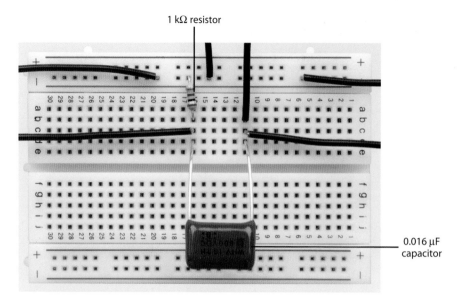

PROJECT 6.1 BREADBOARDED CIRCUIT

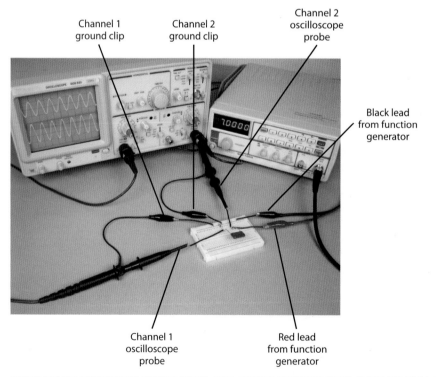

PROJECT 6.1 FUNCTION GENERATOR AND OSCILLOSCOPE ATTACHED TO THE CIRCUIT

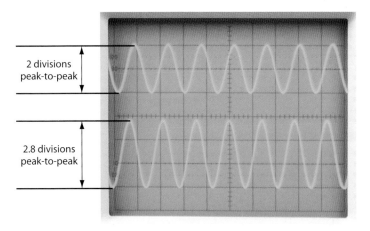

2 divisions peak-to-peak

2.8 divisions peak-to-peak

PROJECT 6.1 INPUT SIGNAL SINE WAVE (UPPER) AND OUTPUT SIGNAL SINE WAVE (LOWER)

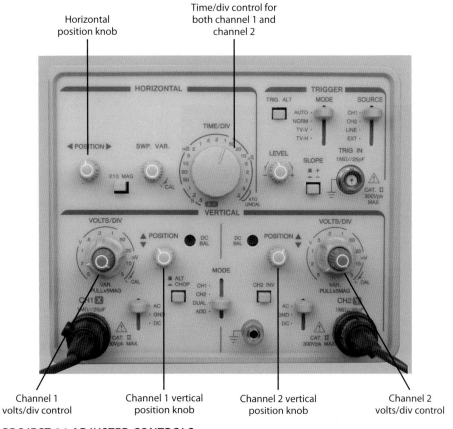

Horizontal position knob

Time/div control for both channel 1 and channel 2

Channel 1 volts/div control

Channel 1 vertical position knob

Channel 2 vertical position knob

Channel 2 volts/div control

PROJECT 6.1 ADJUSTED CONTROLS

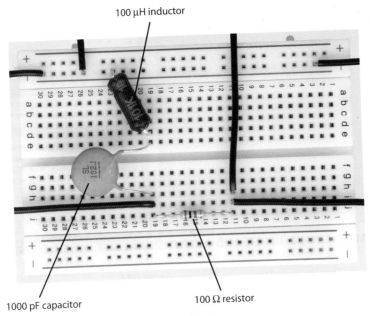

100 µH inductor

1000 pF capacitor

100 Ω resistor

PROJECT 7.1 BREADBOARDED CIRCUIT

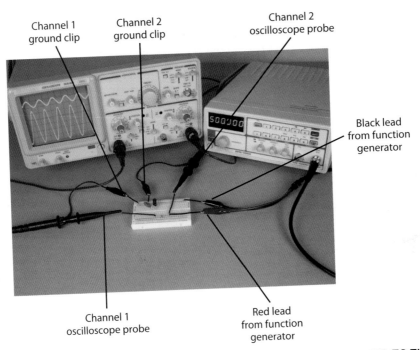

Channel 1
ground clip

Channel 2
ground clip

Channel 2
oscilloscope probe

Black lead
from function
generator

Channel 1
oscilloscope probe

Red lead
from function
generator

PROJECT 7.1 FUNCTION GENERATOR AND OSCILLOSCOPE ATTACHED TO THE CIRCUIT

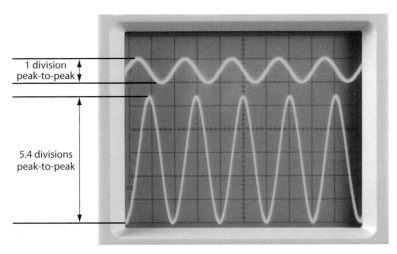

1 division peak-to-peak

5.4 divisions peak-to-peak

PROJECT 7.1 INPUT SIGNAL SINE WAVE (UPPER) AND OUTPUT SIGNAL SINE WAVE (LOWER)

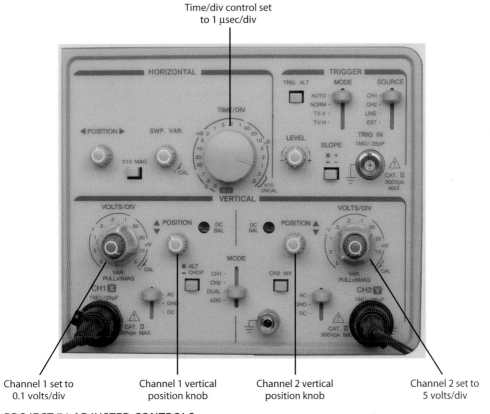

Time/div control set to 1 μsec/div

Channel 1 set to 0.1 volts/div

Channel 1 vertical position knob

Channel 2 vertical position knob

Channel 2 set to 5 volts/div

PROJECT 7.1 ADJUSTED CONTROLS

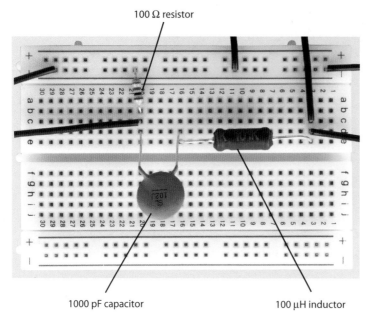

100 Ω resistor

1000 pF capacitor   100 µH inductor

PROJECT 7.2 BREADBOARDED CIRCUIT

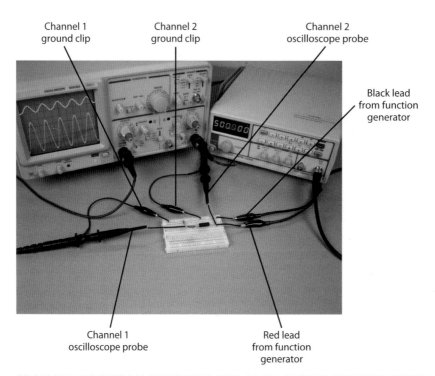

Channel 1
ground clip

Channel 2
ground clip

Channel 2
oscilloscope probe

Black lead
from function
generator

Channel 1
oscilloscope probe

Red lead
from function
generator

PROJECT 7.2 FUNCTION GENERATOR AND OSCILLOSCOPE ATTACHED TO THE CIRCUIT

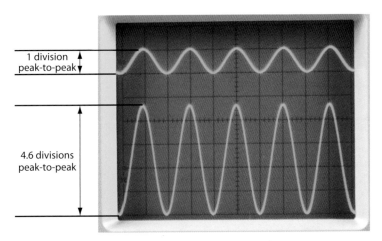

1 division
peak-to-peak

4.6 divisions
peak-to-peak

PROJECT 7.2 INPUT SIGNAL SINE WAVE (UPPER) AND OUTPUT SIGNAL SINE WAVE (LOWER)

Time/div control set
to 1μsec/div

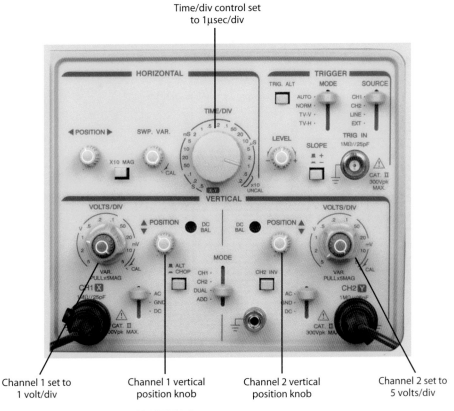

Channel 1 set to
1 volt/div

Channel 1 vertical
position knob

Channel 2 vertical
position knob

Channel 2 set to
5 volts/div

PROJECT 7.2 ADJUSTED CONTROLS

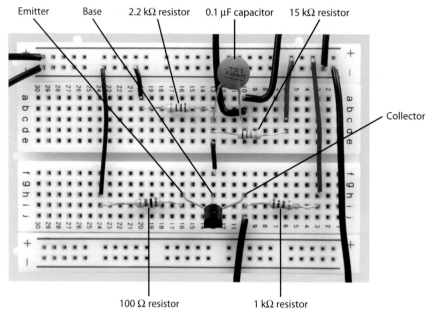

Emitter   Base   2.2 kΩ resistor   0.1 µF capacitor   15 kΩ resistor

Collector

100 Ω resistor   1 kΩ resistor

PROJECT 8.1 BREADBOARDED CIRCUIT #1

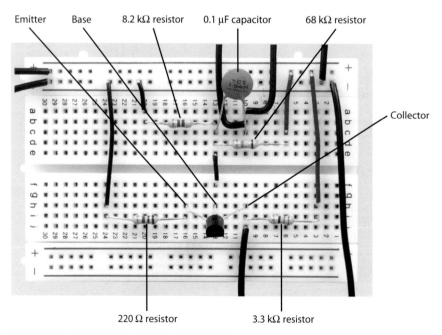

Emitter   Base   8.2 kΩ resistor   0.1 µF capacitor   68 kΩ resistor

Collector

220 Ω resistor   3.3 kΩ resistor

PROJECT 8.1 BREADBOARDED CIRCUIT #3

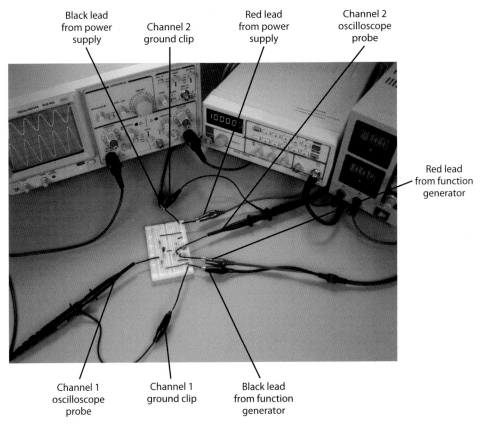

Black lead from power supply   Channel 2 ground clip   Red lead from power supply   Channel 2 oscilloscope probe

Red lead from function generator

Channel 1 oscilloscope probe   Channel 1 ground clip   Black lead from function generator

PROJECT 8.1 FUNCTION GENERATOR AND OSCILLOSCOPE ATTACHED TO THE CIRCUIT

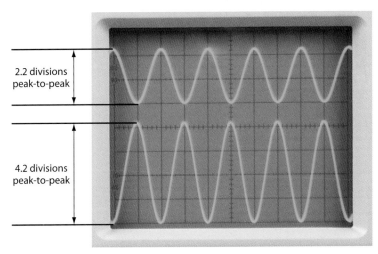

2.2 divisions peak-to-peak

4.2 divisions peak-to-peak

PROJECT 8.1 INPUT SIGNAL SINE WAVE (UPPER) AND OUTPUT SIGNAL SINE WAVE (LOWER)

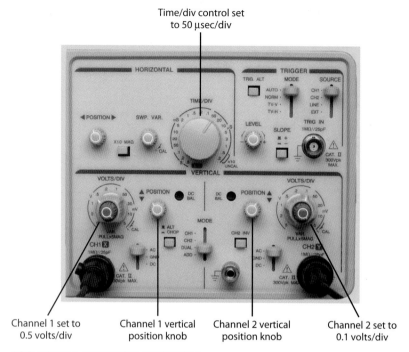

Time/div control set
to 50 μsec/div

Channel 1 set to
0.5 volts/div

Channel 1 vertical
position knob

Channel 2 vertical
position knob

Channel 2 set to
0.1 volts/div

PROJECT 8.1 ADJUSTED CONTROLS

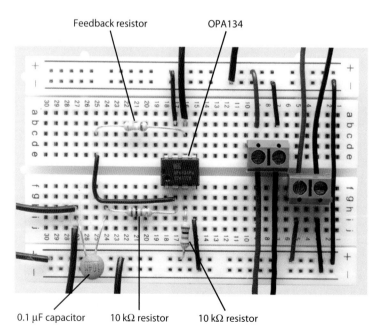

Feedback resistor

OPA134

0.1 μF capacitor

10 kΩ resistor

10 kΩ resistor

PROJECT 8.2 BREADBOARDED CIRCUIT

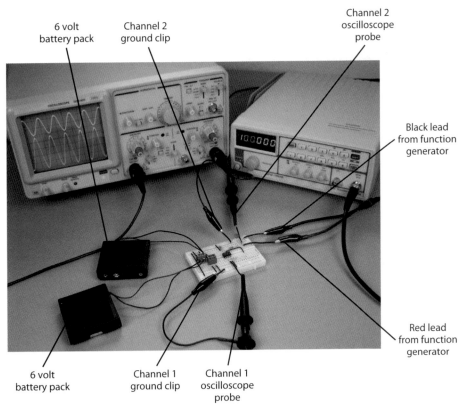

6 volt battery pack

Channel 2 ground clip

Channel 2 oscilloscope probe

Black lead from function generator

Red lead from function generator

6 volt battery pack

Channel 1 ground clip

Channel 1 oscilloscope probe

PROJECT 8.2 FUNCTION GENERATOR AND OSCILLOSCOPE ATTACHED TO THE CIRCUIT

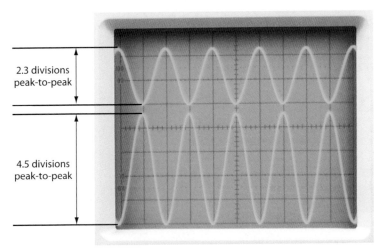

2.3 divisions peak-to-peak

4.5 divisions peak-to-peak

PROJECT 8.2 INPUT SIGNAL SINE WAVE (UPPER) AND OUTPUT SIGNAL SINE WAVE (LOWER)

Time/div control
set to 50 μsec/div

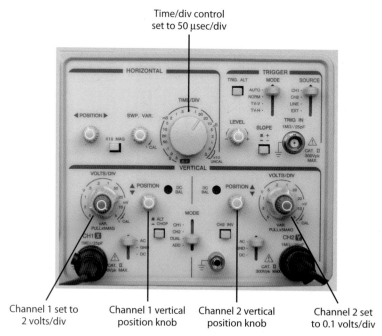

Channel 1 set to
2 volts/div

Channel 1 vertical
position knob

Channel 2 vertical
position knob

Channel 2 set
to 0.1 volts/div

PROJECT 8.2 ADJUSTED CONTROLS

510 Ω
resistor

8.2 kΩ
resistor

1 μF
capacitor

82 kΩ
resistor

10 kΩ
resistor

0.1 μF
capacitor

0.22 μF
capacitor

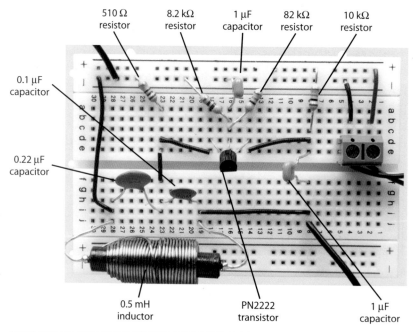

0.5 mH
inductor

PN2222
transistor

1 μF
capacitor

PROJECT 9.1 BREADBOARDED COLPITTS OSCILLATOR WITH FEEDBACK TO THE EMITTER (CIRCUIT # 1)

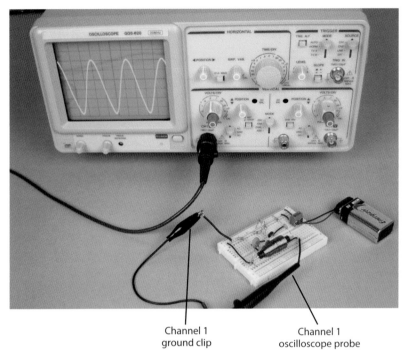

Channel 1
ground clip

Channel 1
oscilloscope probe

PROJECT 9.1 OSCILLOSCOPE ATTACHED TO THE CIRCUIT

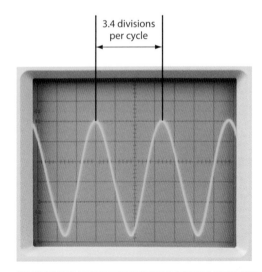

3.4 divisions
per cycle

PROJECT 9.1 THE SINE WAVE GENERATED BY OSCILLATOR CIRCUIT #1

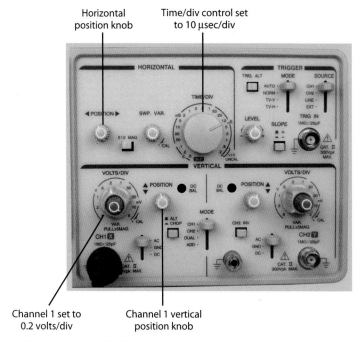

Horizontal position knob

Time/div control set to 10 µsec/div

Channel 1 set to 0.2 volts/div

Channel 1 vertical position knob

PROJECT 9.1 ADJUSTED CONTROLS

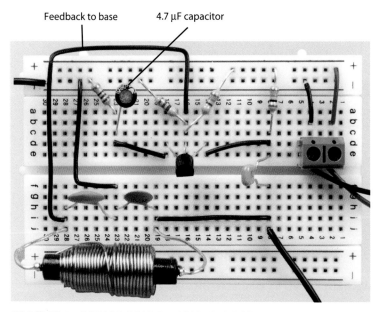

Feedback to base

4.7 µF capacitor

PROJECT 9.1 BREADBOARDED COLPITTS OSCILLATOR WITH FEEDBACK TO THE BASE (CIRCUIT # 2)

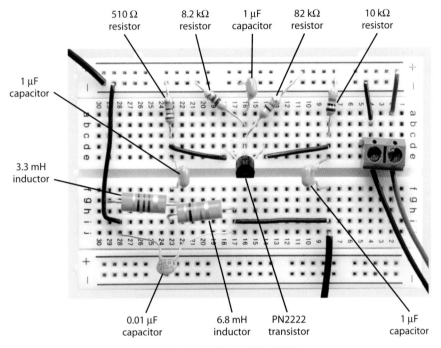

510 Ω resistor

8.2 kΩ resistor

1 μF capacitor

82 kΩ resistor

10 kΩ resistor

1 μF capacitor

3.3 mH inductor

0.01 μF capacitor

6.8 mH inductor

PN2222 transistor

1 μF capacitor

PROJECT 9.2 BREADBOARDED HARTLEY OSCILLATOR

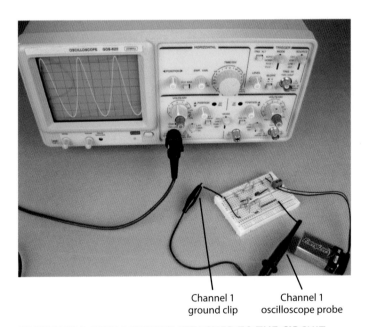

Channel 1 ground clip

Channel 1 oscilloscope probe

PROJECT 9.2 OSCILLOSCOPE ATTACHED TO THE CIRCUIT

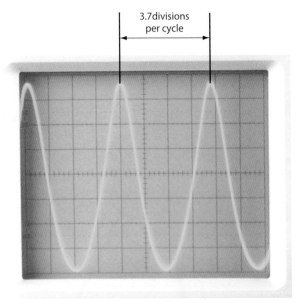

3.7divisions per cycle

PROJECT 9.2 THE SINE WAVE GENERATED BY THE HARTLEY OSCILLATOR

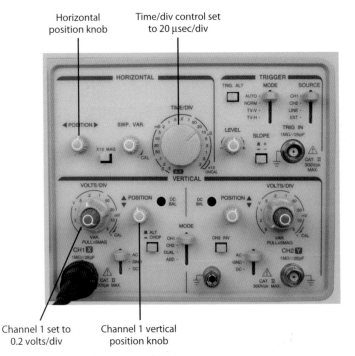

Horizontal position knob

Time/div control set to 20 μsec/div

Channel 1 set to 0.2 volts/div

Channel 1 vertical position knob

PROJECT 9.2 ADJUSTED CONTROLS

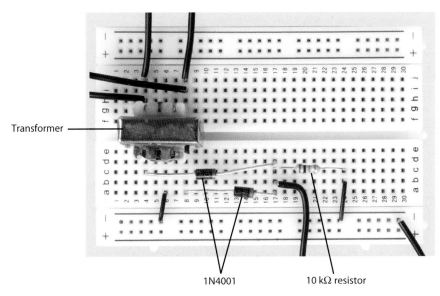

Transformer

1N4001   10 kΩ resistor

PROJECT 11.1 BREADBOARDED FULL- WAVE RECTIFIER (CIRCUIT # 1)

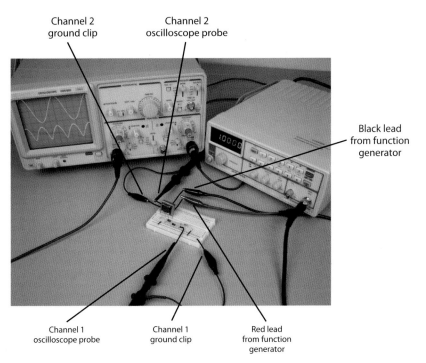

Channel 2
ground clip

Channel 2
oscilloscope probe

Black lead
from function
generator

Channel 1
oscilloscope probe

Channel 1
ground clip

Red lead
from function
generator

PROJECT 11.1 FUNCTION GENERATOR AND OSCILLOSCOPE ATTACHED TO THE CIRCUIT

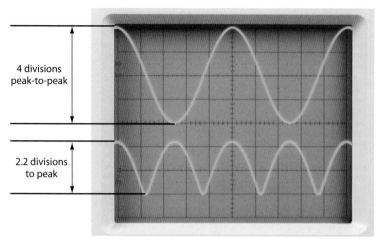

4 divisions
peak-to-peak

2.2 divisions
to peak

PROJECT 11.1 INPUT SIGNAL SINE WAVE (UPPER) AND PULSED DC OUTPUT SIGNAL (LOWER)

Time/div control set
to 0.2 msec/div

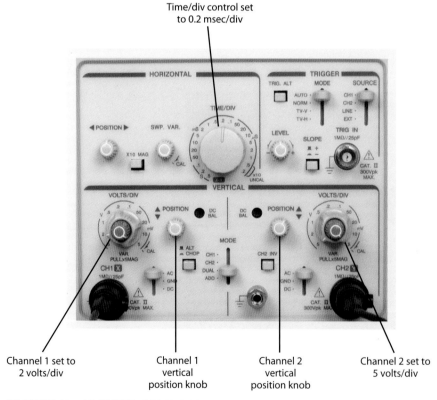

Channel 1 set to
2 volts/div

Channel 1
vertical
position knob

Channel 2
vertical
position knob

Channel 2 set to
5 volts/div

PROJECT 11.1 ADJUSTED CONTROLS

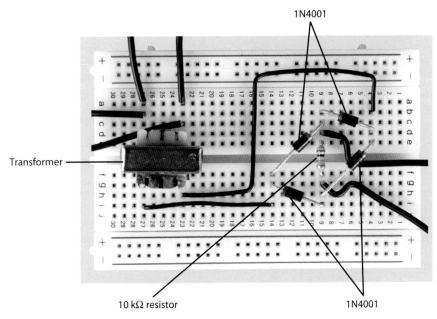

PROJECT 11.1 BREADBOARDED BRIDGE RECTIFIER (CIRCUIT #2)

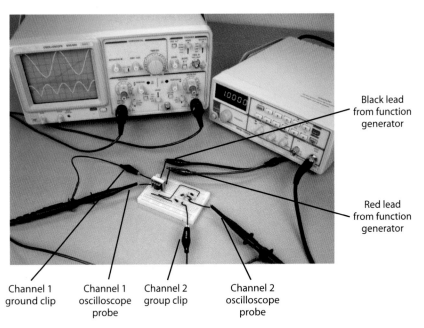

PROJECT 11.1 FUNCTION GENERATOR AND OSCILLOSCOPE ATTACHED TO THE CIRCUIT #2

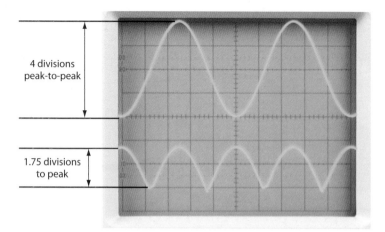

4 divisions
peak-to-peak

1.75 divisions
to peak

**PROJECT 11.1 INPUT SIGNAL SINE WAVE (UPPER) AND PULSED DC OUTPUT SIGNAL (LOWER)
FOR CIRCUIT #2**

Time/div control set
to 0.2 msec/div

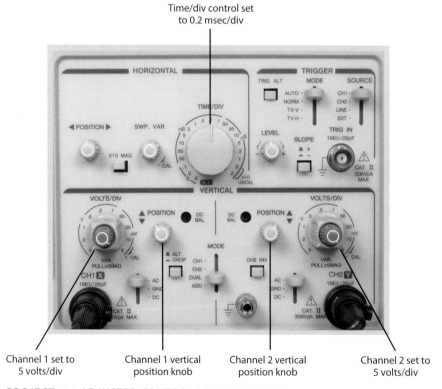

Channel 1 set to
5 volts/div

Channel 1 vertical
position knob

Channel 2 vertical
position knob

Channel 2 set to
5 volts/div

PROJECT 11.1 ADJUSTED CONTROLS FOR CIRCUIT #2

7

Resonant Circuits

You have seen how the inductor and the capacitor each present an opposition to the flow of an AC current, and how the magnitude of this reactance depends upon the frequency of the applied signal.

When inductors and capacitors are used together in a circuit (referred to as an LC circuit), a useful phenomenon called *resonance* occurs. Resonance is the frequency at which the reactance of the capacitor and the inductor is equal.

In this chapter, you learn about some of the properties of resonant circuits, and concentrate on those properties that lead to the study of oscillators (which is touched upon in the last few problems in this chapter and covered in more depth in Chapter 9, "Oscillators").

After completing this chapter, you will be able to do the following:

- Find the impedance of a series LC circuit.
- Calculate the series LC circuit's resonant frequency.
- Sketch a graph of the series LC circuit's output voltage.
- Find the impedance of a parallel LC circuit.
- Calculate the parallel LC circuit's resonant frequency.
- Sketch a graph of the parallel LC circuit's output voltage.
- Calculate the bandwidth and the quality factor (Q) of simple series and parallel LC circuits.
- Calculate the frequency of an oscillator.

THE CAPACITOR AND INDUCTOR IN SERIES

1 Many electronic circuits contain a capacitor and an inductor placed in series, as shown in Figure 7.1.

FIGURE 7.1

You can combine a capacitor and an inductor in series with a resistor to form voltage divider circuits, such as the two circuits shown in Figure 7.2. A circuit that contains resistance (R), inductance (L), and capacitance (C) is referred to as an RLC circuit. Although the order of the capacitor and inductor differs in the two circuits shown in Figure 7.2, they have the same effect on electrical signals.

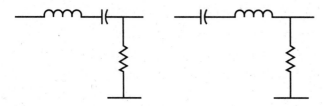

FIGURE 7.2

To simplify your calculations in the next few problems, you can assume that the small DC resistance of the inductor is much less than the resistance of the resistor R, and you can, therefore, ignore DC resistance in your calculations.

When you apply an AC signal to the circuits in Figure 7.2, both the inductor's and the capacitor's reactance value depends on the frequency.

QUESTIONS

A. What formula would you use to calculate the inductor's reactance? _____

_____

B. What formula would you use to calculate the capacitor's reactance? _____

_____

ANSWERS

A. $X_L = 2\pi fL$

B. $X_C = \dfrac{1}{2\pi fC}$

2 You can calculate the net reactance (X) of a capacitor and inductor in series by using the following formula:

$$X = X_L - X_C$$

You can calculate the impedance of the RLC circuits shown in Figure 7.2 by using the following formula:

$$Z = \sqrt{R^2 + X^2}$$

In the formula, keep in mind that X^2 is $(X_L - X_C)^2$.

Calculate the net reactance and impedance for an RLC series circuit, such as those shown in Figure 7.2, with the following values:

f = 1 kHz

L = 100 mH

C = 1 μF

R = 500 ohms

QUESTIONS

Follow these steps to calculate the following:

A. Find X_L: _____

_____

B. Find X_C: _____

_____

C. Use $X = X_L - X_C$ to find the net reactance: _____

_____

D. Use $Z = \sqrt{X^2 + R^2}$ to find the impedance: _____

_____

ANSWERS

A. $X_L = 628$ ohms

B. $X_C = 160$ ohms

C. $X = 468$ ohms (inductive)

D. $Z = 685$ ohms

3 Calculate the net reactance and impedance for an RLC series circuit, such as those shown in Figure 7.2, using the following values:

$f = 100$ Hz

$L = 0.5$ H

$C = 5$ μF

$R = 8$ ohms

QUESTIONS

Follow the steps outlined in problem 2 to calculate the following parameters:

A. $X_L =$ _____

B. $X_C =$ _____

C. $X =$ _____

D. $Z =$ _____

ANSWERS

A. $X_L = 314$ ohms

B. $X_C = 318$ ohms

C. $X = -4$ ohms (capacitive)

D. $Z = 9$ ohms

By convention, the net reactance is negative when it is capacitive.

4 Calculate the net reactance and impedance for an RLC series circuit, such as those shown in Figure 7.2, using the values in the following questions.

QUESTIONS

A. f = 10 kHz, L = 15 mH, C = 0.01 μF, R = 494 ohms

X = _____

Z = _____

B. f = 2 MHz, L = 8 μH, C = 0.001 μF, R = 15 ohms

X = _____

Z = _____

ANSWERS

A. X = −650 ohms (capacitive), Z = 816 ohms

B. X = 21 ohms (inductive), Z = 25.8 ohms

5 For the circuit shown in Figure 7.3, the output voltage is the voltage drop across the resistor.

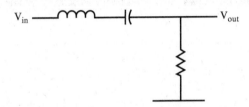

FIGURE 7.3

In problems 1 through 4, the net reactance of the series inductor and capacitor changes as the frequency changes. Therefore, as the frequency changes, the voltage drop across the resistor changes and so does the amplitude of the output voltage V_{out}.

If you plot V_{out} against frequency on a graph for the circuit shown in Figure 7.3, the curve looks like the one shown in Figure 7.4.

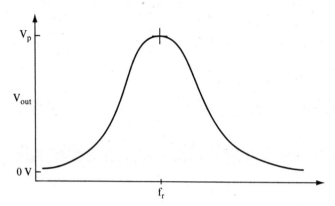

FIGURE 7.4

The maximum output voltage (or peak voltage) shown in this curve, V_p, is slightly less than V_{in}. This slight attenuation of the peak voltage from the input voltage is because of the DC resistance of the inductor.

The output voltage peaks at a frequency, f_r, where the net reactance of the inductor and capacitor in series is at its lowest value. At this frequency, there is little voltage drop across the inductor and capacitor. Therefore, most of the input voltage is applied across the resistor, and the output voltage is at its highest value.

QUESTION

Under ideal conditions, if X_C were 10.6 ohms, what value of X_L results in a net reactance (X) of 0 for the circuit shown in Figure 7.3? _____

_____

ANSWER

$X = X_L - X_C = 0$, therefore:
$X_L = X_C + X = 10.6 \, \Omega + 0 = 10.6 \, \Omega$

6 You can find the frequency at which $X_L - X_C = 0$ by setting the formula for X_L equal to the formula for X_C and solving for f:

$$2\pi fL = \frac{1}{2\pi fC}$$

Therefore,

$$f_r = \frac{1}{2\pi\sqrt{LC}}$$

where f_r is the *resonant frequency* of the circuit.

QUESTION

What effect does the value of the resistance have on the resonant frequency? _____
_____

ANSWER

It has no effect at all.

7 Calculate the resonant frequency for the circuit shown in Figure 7.3 using the capacitor and inductor values given in the following questions.

QUESTIONS

A. $C = 1\ \mu F$, $L = 1\ mH$_____

   $f_r =$ _____

B. $C = 16\ \mu F$, $L = 1.6\ mH$_____

   $f_r =$ _____

ANSWERS

A. $f_r = \dfrac{1}{2\pi\sqrt{1 \times 10^{-3} \times 1 \times 10^{-6}}} = 5.0\,kHz$

B. $f_r = \dfrac{1}{2\pi\sqrt{16 \times 10^{-6} \times 1.6 \times 10^{-3}}} = 1\,kHz$

8 Calculate the resonant frequency for the circuit shown in Figure 7.3 using the capacitor and inductor values given in the following questions.

QUESTIONS

   A. $C = 0.1\ \mu F, L = 1\ mH$ _____

      $f_r =$ _____

   B. $C = 1\ \mu F, L = 2\ mH$ _____

      $f_r =$ _____

ANSWERS

   A. $f_r = 16\ kHz$

   B. $f_r = 3.6\ kHz$

---

9 For the RLC circuit shown in Figure 7.5, the output voltage is the voltage drop across the capacitor and inductor.

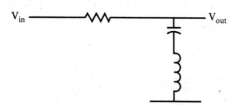

FIGURE 7.5

If V_{out} is plotted on a graph against the frequency for the circuit shown in Figure 7.5, the curve looks like that shown in Figure 7.6.

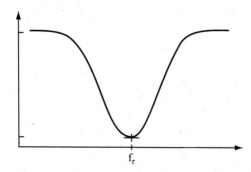

FIGURE 7.6

The output voltage drops to its minimum value at the resonant frequency for the circuit, which you can calculate with the formula provided in problem 6. At the resonant frequency, the net reactance of the inductor and capacitor in series is at a minimum. Therefore, there is little voltage drop across the inductor and capacitor, and the output voltage is at its minimum value. A circuit with this type of output (such as the circuit shown in Figure 7.5) is called a *notch filter*, or *band-reject filter*.

QUESTION

What would you expect the minimum output voltage to be? _____

_____

ANSWER

0 volts, or close to it

PROJECT 7.1: The Notch Filter

OBJECTIVE

The objective of this project is to determine how V_{out} changes as the frequency of the input signal changes for a notch filter.

GENERAL INSTRUCTIONS

After the circuit is set up, you measure V_{out} for each frequency. You can also generate a graph to show the relationship between the output voltage and the input frequency.

Parts List

You need the following equipment and supplies:

- ❑ One 100 Ω, 0.25-watt resistor.
- ❑ One 1000 pF capacitor. (1000 pF is also sometimes stated by suppliers as 0.001 μF.)
- ❑ One 100 μH inductor. You'll often find inductors that use a numerical code to indicate the value of the inductor. The first two numbers in this code are the first

and second significant digits of the inductance value. The third number is the multiplier, and the units are μH. Therefore, an inductor marked with 101 has a value of 100 μH.)

❑ One function generator.

❑ One oscilloscope.

❑ One breadboard.

STEP-BY-STEP INSTRUCTIONS

Set up the circuit shown in Figure 7.7. If you have some experience in building circuits, this schematic (along with the previous parts list) should provide all the information you need to build the circuit. If you need a bit more help building the circuit, look at the photos of the completed circuit in the "Expected Results" section.

Carefully check your circuit against the diagram.

After you check your circuit, follow these steps, and record your measurements in the blank table following the steps.

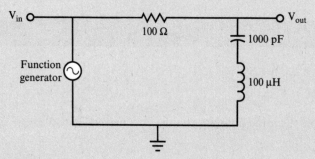

FIGURE 7.7

1. Connect the oscilloscope probe for channel 2 to a jumper wire connected to V_{in}, and connect the ground clip to a jumper wire attached to the ground bus.

2. Connect the oscilloscope probe for channel 1 to a jumper wire connected to V_{out}, and connect the ground clip to a jumper wire attached to the ground bus.

3. Set the function generator to generate a 5 V_{pp}, 100 kHz sine wave.

4. Measure and record V_{out}.

5. Adjust the function generator to the frequency shown in the next row of the table (labeled 150 kHz in this instance). Each time you change the frequency, check V_{in} and adjust the amplitude knob on the function generator to maintain V_{in} at 5 V_{pp} if needed. (If you leave the amplitude knob in one position, the voltage of the signal provided by the function generator will change as the net reactance of the circuit changes.)

6. Measure and record V_{out}.

7. Repeat steps 5 and 6 for the remaining values until you have recorded V_{out} in all rows of the table.

f_{in} (kHz)	V_{out} (volts)
100	
150	
200	
250	
300	
350	
400	
450	
500	
550	
600	
650	
700	
750	
800	
850	
900	

8. In the blank graph shown in Figure 7.8, plot V_{out} vs f_{in} with the voltage on the vertical axis and the frequency on the X axis. The curve should have the same shape as the curve shown in Figure 7.6.

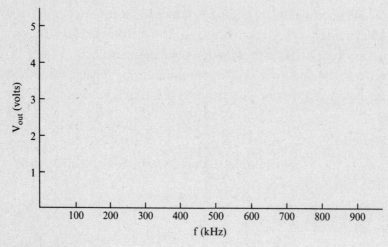

FIGURE 7.8

EXPECTED RESULTS

Figure 7.9 shows the breadboarded circuit for this project.

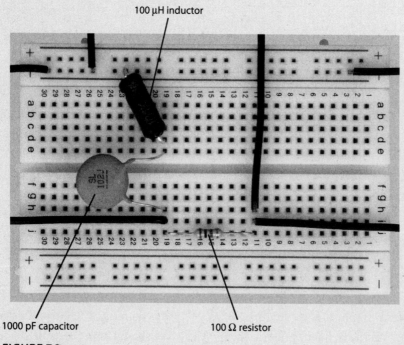

100 μH inductor

1000 pF capacitor   100 Ω resistor

FIGURE 7.9

Figure 7.10 shows a function generator and oscilloscope attached to the circuit.

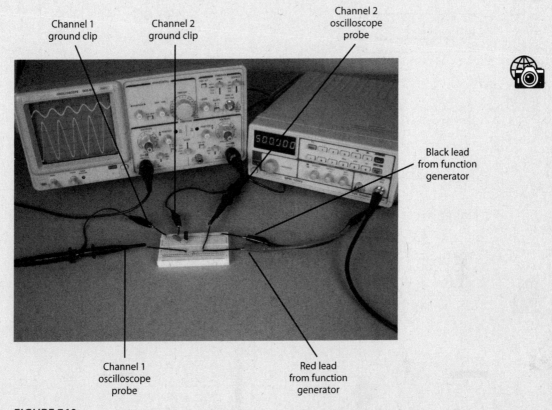

Channel 1 ground clip

Channel 2 ground clip

Channel 2 oscilloscope probe

Black lead from function generator

Channel 1 oscilloscope probe

Red lead from function generator

FIGURE 7.10

The input signal is represented by the upper sine wave shown in Figure 7.11, and the output signal is represented by the lower sine wave. Read the number of divisions for the peak-to-peak output sine wave, and multiply it by the corresponding VOLTS/DIV setting to determine V_{out}.

As you set f_{in} to a new value on the function generator, you may also need to adjust the TIME/DIV control, the VOLTS/DIV control, and vertical POSITION controls on the oscilloscope. The controls shown in Figure 7.12 are adjusted to measure V_{out} when f_{in} = 500 kHz.

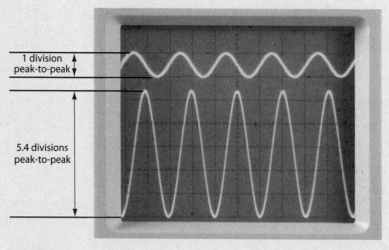

1 division
peak-to-peak

5.4 divisions
peak-to-peak

FIGURE 7.11

Time/div control set
to 1 μsec/div

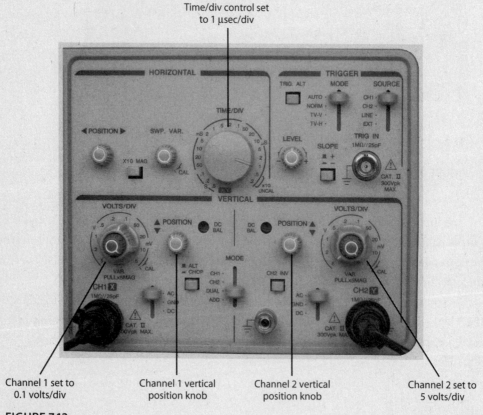

Channel 1 set to
0.1 volts/div

Channel 1 vertical
position knob

Channel 2 vertical
position knob

Channel 2 set to
5 volts/div

FIGURE 7.12

Your values should be close to those shown in the following table, and the curve should be similar to that shown in Figure 7.13.

f_{in} (kHz)	V_{out} (volts)
100	4.9
150	4.9
200	4.9
250	4.8
300	4.7
350	4.4
400	4.0
450	2.8
500	0.5
550	2.3
600	3.6
650	4.2
700	4.4
750	4.7
800	4.7
850	4.8
900	4.8

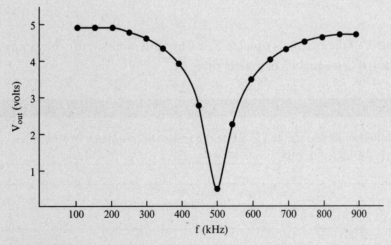

FIGURE 7.13

You may want to take extra data points at frequencies near the minimum V_{out} to help you determine the precise frequency at which the minimum V_{out} occurs. When extra data points are added to this graph, for example, the minimum V_{out} occurs at a frequency of 505 kHz, which is close to the calculated resonance frequency of 503 kHz for this circuit.

10 You can connect the capacitor and inductor in parallel, as shown in Figure 7.14.

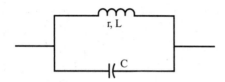

FIGURE 7.14

You can calculate the resonance frequency of this circuit using the following formula:

$$f_r = \frac{1}{2\pi\sqrt{LC}}\sqrt{1 - \frac{r^2C}{L}}$$

In this formula, r is the DC resistance of the inductor. However, if the reactance of the inductor is equal to, or more than, 10 times the DC resistance of the inductor, you can use the following simpler formula. This is the same formula that you used in problems 7 and 8 for the series circuit.

$$f_r = \frac{1}{2\pi\sqrt{LC}}$$

Q, the quality factor of the circuit, is equal to X_L/r. Therefore, you can use this simple equation to calculate f_r if Q is equal to, or greater than, 10.

QUESTIONS

A. Which formula should you use to calculate the resonant frequency of a parallel circuit if the Q of the coil is 20? _____

_____

B. If the Q is 8? _____

_____

ANSWERS

A. $f_r = \dfrac{1}{2\pi\sqrt{LC}}$

B. $f_r = \dfrac{1}{2\pi\sqrt{LC}}\sqrt{1 - \dfrac{r^2C}{L}}$

NOTE Here is another version of the resonance frequency formula that is help-ful when Q is known:

$$f_r = \frac{1}{2\pi\sqrt{LC}}\sqrt{\frac{Q^2}{1+Q^2}}$$

11 You can calculate the total opposition (impedance) of an inductor and capacitor con-nected in parallel to the flow of current by using the following formulas for a circuit at resonance:

$Z_p = Q^2r$, if Q is equal to or greater than 10

$Z_p = \dfrac{L}{rC}$, for any value of Q

At resonance, the impedance of an inductor and capacitor in parallel is at its maximum.

You can use an inductor and capacitor in parallel in a voltage divider circuit, as shown in Figure 7.15.

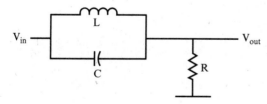

FIGURE 7.15

If V_{out} is plotted against frequency on a graph for the circuit shown in Figure 7.15, the curve looks like that shown in Figure 7.16.

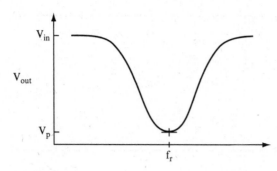

FIGURE 7.16

QUESTIONS

A. What would be the total impedance formula for the voltage divider circuit at resonance? _____

_____

B. What is the frequency called at the point where the curve is at its lowest point? ____

_____

C. Why is the output voltage at a minimum value at resonance? _____

_____

ANSWERS

A. $Z_T = Z_p + R$

NOTE   The relationship shown by this formula is true only at resonance. At all other frequencies, Z_T is a complicated formula or calculation found by considering a series r, L circuit in parallel with a capacitor.

B. The parallel resonant frequency.

C. The output voltage is at its lowest value at the resonant frequency. This is because the impedance of the parallel resonant circuit is at its highest value at this frequency.

12 For the circuit shown in Figure 7.17, the output voltage equals the voltage drop across the inductor and capacitor.

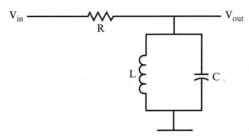

FIGURE 7.17

If V_{out} is plotted on a graph against frequency for the circuit shown in Figure 7.17, the curve looks like that shown in Figure 7.18. At the resonance frequency, the impedance of the parallel inductor and capacitor is at its maximum value. Therefore, the voltage drop across the parallel inductor and capacitor (which is also the output voltage) is at its maximum value.

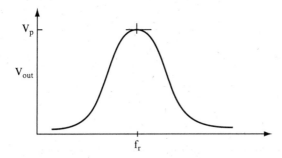

FIGURE 7.18

QUESTION

What formula would you use to calculate the resonant frequency? _____

_____

ANSWER

$$f_r = \frac{1}{2\pi\sqrt{LC}} \text{ if Q is equal to or greater than 10}$$

$$f_r = \frac{1}{2\pi\sqrt{LC}}\sqrt{1 - \frac{r^2C}{L}} \text{ if Q is less than 10}$$

13 Find the resonant frequency in these two examples, where the capacitor and the inductor are in parallel. (Q is greater than 10.)

QUESTIONS

A. L = 5 mH, C = 5 µF_____

f_r = _____

_____

B. L = 1 mH, C = 10 µF_____

f_r = _____

_____

ANSWERS

A. f_r = 1 kHz (approximately)

B. f_r = 1600 Hz (approximately)

THE OUTPUT CURVE

14 Now it's time to look at the output curve in a little more detail. Take a look at the curve shown in Figure 7.19 for an example.

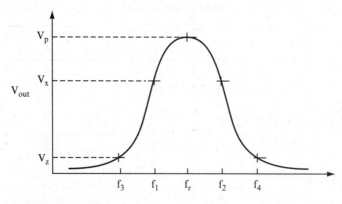

FIGURE 7.19

An input signal at the resonant frequency, f_r, passes through a circuit with minimum attenuation, and with its output voltage equal to the peak output voltage, V_p, shown on this curve.

The two frequencies f_1 and f_2 are "passed" almost as well as f_r is passed. That is, signals at those frequencies have a high output voltage, almost as high as the output of a signal at f_r. The graph shows this voltage as V_x.

Signals at frequencies f_3 and f_4 have a low output voltage.

These two frequencies are not passed but are said to be *blocked* or *rejected* by the circuit. This output voltage is shown on the graph as V_z.

The output or frequency response curve for a resonant circuit (series or parallel) has a symmetrical shape for a high value of Q. You can make the assumption that the output curve is symmetrical when Q is greater than 10.

QUESTIONS

A. What is meant by a frequency that is passed? _____

_____

B. Why are f_1 and f_2 passed almost as well as f_r? _____

_____

C. What is meant by a frequency that is blocked? _____

_____

D. Which frequencies shown on the previous output curve are blocked? _____

_____

E. Does the output curve shown appear to be symmetrical? What does this mean for the circuit?_____

_____

ANSWERS

A. It appears at the output with minimum attenuation.

B. Because their frequencies are close to f_r.

C. It has a low output voltage.

D. f_3 and f_4 (as well as all frequencies below f_3 and above f_4).

E. It does appear to be symmetrical. This means that the coil has a Q greater than 10.

15 Somewhere between f_r and f_3, and between f_r and f_4, there is a point at which frequencies are said to be either passed or reduced to such a level that they are effectively blocked. The dividing line is at the level at which the power output of the circuit is half as much as the power output at peak value. This happens to occur at a level that is 0.707, or 70.7 percent of the peak value.

For the output curve shown in problem 14, this occurs at a voltage level of 0.707 V_p. The two corresponding frequencies taken from the graph are called the *half power frequencies* or *half power points*. These are common expressions used in the design of resonant circuits and frequency response graphs.

If a certain frequency results in an output voltage that is equal to or greater than the half power point, it is said to be passed or accepted by the circuit. If it is lower than the half power point, it is said to be blocked or rejected by the circuit.

QUESTION

Suppose V_p = 10 volts. What is the minimum voltage level of all frequencies that are passed by the circuit? _____

ANSWER

V = 10 volts × 0.707 = 7.07 volts

(If a frequency has an output voltage above 7.07 volts, you would say it is passed by the circuit.)

16 Assume the output voltage at the resonant frequency in a circuit is 5 volts. Another frequency has an output of 3.3 volts.

QUESTION

Is this second frequency passed or blocked by the circuit? _____

ANSWER

V = V_p × 0.707 = 5 × 0.707 = 3.535 volts
3.3 volts is less than 3.535 volts, so this frequency is blocked.

17 In these examples, find the voltage level at the half power points.

QUESTIONS

A. V_p = 20 volts _____

_____

B. V_p = 100 volts _____

_____

C. V_p = 3.2 volts _____

_____

ANSWERS

A. 14.14 volts

B. 70.70 volts

C. 2.262 volts

18 Although this discussion started off by talking about the resonance frequency, a few other frequencies have been introduced. At this point, the discussion is dealing with a band or a range of frequencies.

Two frequencies correspond to the half power points on the curve. Assume these frequencies are f_1 and f_2. The difference you find when you subtract f_1 from f_2 is important because this gives the range of frequencies that are passed by the circuit. This range is called the *bandwidth* of the circuit and can be calculated using the following equation:

$$BW = f_2 - f_1$$

All frequencies within the bandwidth are passed by the circuit, whereas all frequencies outside the bandwidth are blocked. A circuit with this type of output (such as the circuit shown in Figure 7.17) is referred to as a *bandpass filter*.

QUESTION

Indicate which of the following pairs of values represent a wider range of frequencies, or, in other words, the wider bandwidth.

A. f_2 = 200 Hz, f_1 = 100 Hz

B. f_2 = 20 Hz, f_1 = 10 Hz

When playing a radio, you listen to one station at a time, not to the adjacent stations on the dial. Thus, your radio tuner must have a narrow bandwidth so that it can select only the frequency of that one station.

The amplifiers in a television set, however, must pass frequencies from 30 Hz up to approximately 4.5 MHz, which requires a wider bandwidth. The application or use to which you'll put a circuit determines the bandwidth that you should design the circuit to provide.

19 The output curve for a circuit that passes a band of frequencies around the resonance frequency (such as the curve shown in Figure 7.20) was discussed in the last few problems.

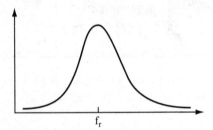

f_r

FIGURE 7.20

The same principles and equations apply to the output curve for a circuit that blocks a band of frequencies around the resonance frequency, as is the case with the curve shown in Figure 7.21.

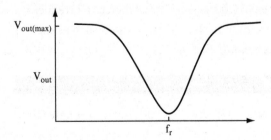

$V_{out(max)}$

V_{out}

f_r

FIGURE 7.21

QUESTIONS

A. What points on the curve shown in Figure 7.21 would you use to determine the circuit's bandwidth? _____

_____

B. Would the output voltage at the resonant frequency be above or below these points? _____

_____

ANSWERS

A. The half power points ($0.707\ V_{out(max)}$).

B. The output voltage at the resonant frequency is the minimum point on the curve, which is below the level for the half power points.

PROJECT 7.2: The Band Pass Filter

OBJECTIVE

The objective of this project is to determine how V_{out} changes as the frequency of the input signal changes for a bandpass filter.

GENERAL INSTRUCTIONS

When the circuit is set up, you measure V_{out} for each frequency. You also generate a graph to show the relationship between the output voltage and the input frequency.

Parts List

You need the following equipment and supplies:

- ❑ One 100 Ω, 0.25-watt resistor.
- ❑ One 1000 pF capacitor. (1000 pF is also sometimes stated by suppliers as 0.001 μF.)
- ❑ One 100 μH inductor.

❑ One function generator.

❑ One oscilloscope.

❑ One breadboard.

STEP-BY-STEP INSTRUCTIONS

Set up the circuit shown in Figure 7.22. If you have some experience in building circuits, this schematic (along with the previous parts list) should provide all the information you need to build the circuit. If you need a bit more help building the circuit, look at the photos of the completed circuit in the "Expected Results" section.

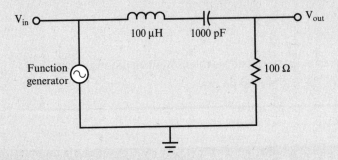

FIGURE 7.22

Carefully check your circuit against the diagram.

After you check your circuit, follow these steps, and record your measurements in the blank table following the steps.

1. Connect the oscilloscope probe for channel 2 to a jumper wire connected to V_{in}, and connect the ground clip to a jumper wire attached to the ground bus.

2. Connect the oscilloscope probe for channel 1 to a jumper wire connected to V_{out}, and connect the ground clip to a jumper wire attached to the ground bus.

3. Set the function generator to generate a 5 V_{pp}, 100 kHz sine wave.

4. Measure and record V_{out}.

5. Adjust the function generator to the frequency shown in the next row of the table (labeled 150 kHz in this instance). Each time you change the frequency, check V_{in},

and adjust the amplitude knob on the function generator to maintain V_{in} at 5 V_{pp} if needed. (If you leave the amplitude knob in one position, the voltage of the signal provided by the function generator will change as the net reactance of the circuit changes.)

6. Measure and record V_{out}.

7. Repeat steps 5 and 6 until you have recorded V_{out} for the last row of the table.

f_{in} (kHz)	V_{out} (volts)
100	
150	
200	
250	
300	
350	
400	
450	
500	
550	
600	
650	
700	
750	
800	
850	
900	

8. In the blank graph shown in Figure 7.23, plot V_{out} versus f_{in} with the voltage on the vertical axis and the frequency on the X axis. The curve should have the same shape as the curve shown in Figure 7.20.

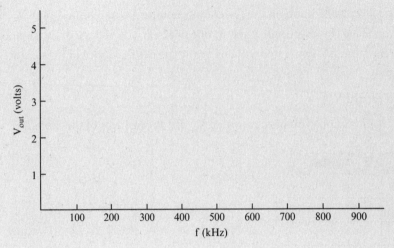

FIGURE 7.23

EXPECTED RESULTS

Figure 7.24 shows the breadboarded circuit for this project.

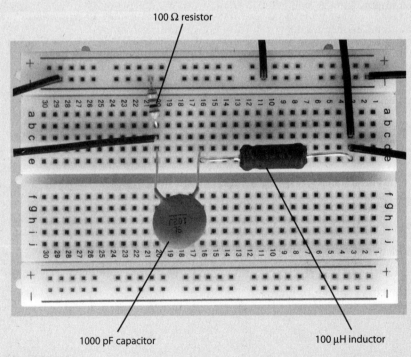

FIGURE 7.24

Figure 7.25 shows a function generator and oscilloscope attached to the circuit.

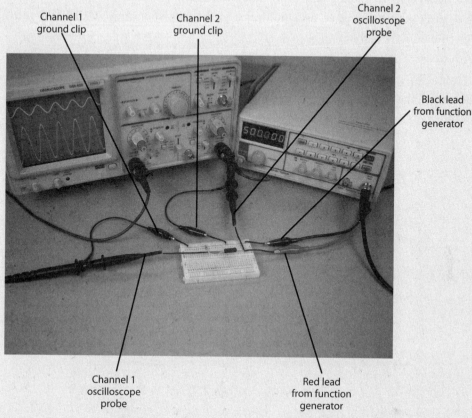

FIGURE 7.25

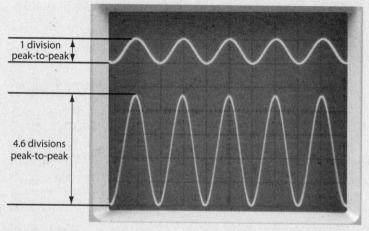

FIGURE 7.26

The input signal is represented by the upper sine wave shown in Figure 7.26, and the output signal is represented by the lower sine wave. Read the number of divisions for the peak-to-peak output sine wave, and multiply it by the corresponding VOLTS/DIV setting to determine V_{out}.

As you set f_{in} to a new value on the function generator, you may also need to adjust the TIME/DIV control, the VOLTS/DIV control, and vertical POSITION controls on the oscilloscope. The controls shown in Figure 7.27 are adjusted to measure V_{out} when f_{in} = 500 kHz.

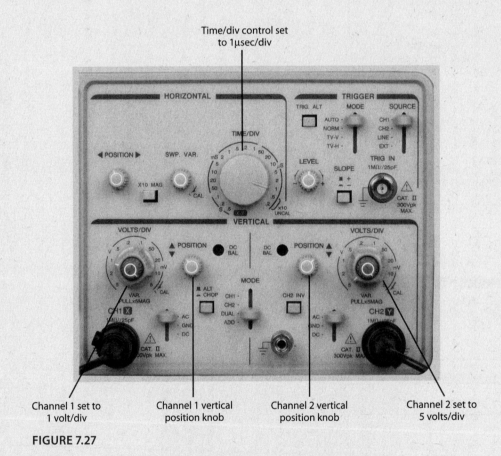

Time/div control set to 1μsec/div

Channel 1 set to 1 volt/div

Channel 1 vertical position knob

Channel 2 vertical position knob

Channel 2 set to 5 volts/div

FIGURE 7.27

Your values should be close to those shown in the following table, and the curve should be similar to that shown Figure 7.28.

f_{in} (kHz)	V_{out} (volts)
100	0.3
150	0.5
200	0.7
250	1.0
300	1.4
350	1.8
400	2.6
450	3.6
500	4.6
550	4.2
600	3.4
650	2.7
700	2.2
750	1.8
800	1.6
850	1.4
900	1.3

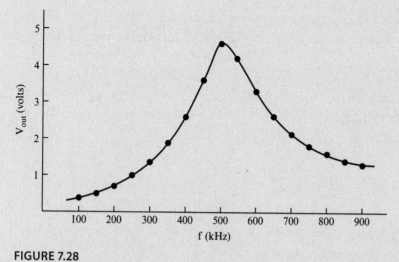

FIGURE 7.28

Because Q = 3.2 (well below 10), the curve for this circuit is not perfectly symmetrical.

20 You can find the bandwidth of a circuit by measuring the frequencies (f_1 and f_2) at which the half power points occur and then using the following formula:

$$BW = f_2 - f_1$$

Or you can calculate the bandwidth of a circuit using this formula:

$$BW = \frac{f_r}{Q}$$

where:

$$Q = \frac{X_L}{R}$$

The formula used to calculate bandwidth indicates that, for two circuits with the same resonant frequency, the circuit with the larger Q will have the smaller bandwidth.

When you calculate Q for a circuit containing a capacitor and inductor in series (such as that shown in Figure 7.29), use the total DC resistance—the sum of the DC resistance (r) of the inductor and the value of the resistor (R)—to calculate Q.

When you calculate Q for a circuit containing an inductor and capacitor in parallel, as with the circuit shown in Figure 7.30, you do not include the value of the resistor (R) in the calculation. The only resistance you use in the calculation is the DC resistance (r) of the inductor.

When you calculate Q for a circuit containing an inductor, a capacitor, and a resistor in parallel (as with the two circuits shown in Figure 7.31), include the value of the resistor (R) in the calculation.

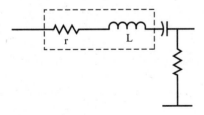

FIGURE 7.29

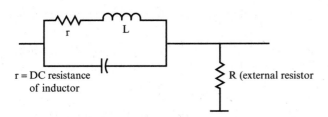

r = DC resistance
    of inductor

R (external resistor

FIGURE 7.30

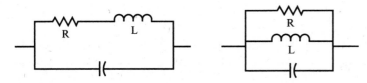

FIGURE 7.31

QUESTIONS

For the circuit shown in Figure 7.32, all the component values are provided in the diagram. Find f_r, Q, and BW.

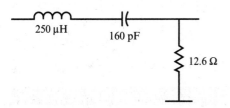

250 μH

160 pF

12.6 Ω

FIGURE 7.32

A. f_r = _____

_____

B. Q = _____

_____

C. BW = _____

_____

ANSWERS

A. $f_r = \dfrac{1}{2\pi\sqrt{LC}} = \dfrac{1}{2\pi\sqrt{250 \times 10^{-6} \times 160 \times 10^{-12}}} = 796\,\text{kHZ}$

B. $Q = \dfrac{X_L}{R} = \dfrac{2\pi fL}{R} = \dfrac{2\pi \times 796\,\text{kHz} \times 250\,\mu\text{H}}{12.6\,\Omega} = 99.2$

C. $BW = \dfrac{f_r}{Q} = \dfrac{796\,\text{kHz}}{99.2} = 8\,\text{kHz}$

21 Use the circuit and component values shown in Figure 7.33 to answer the following questions.

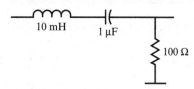

10 mH   1 μF   100 Ω

FIGURE 7.33

QUESTIONS

Find f_r, Q, and BW. Then, on a separate sheet of paper, draw an output curve showing the range of frequencies that are passed and blocked.

$f_r = $ _____

_____

$Q = $ _____

_____

$BW = $ _____

_____

ANSWERS

$f_r = 1590$ Hz; $Q = 1$; BW $= 1590$ Hz
The output curve is shown in Figure 7.34.

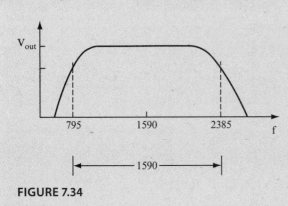

FIGURE 7.34

22 Use the circuit and component values shown in Figure 7.35 to answer the following questions.

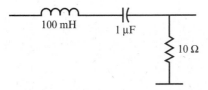

FIGURE 7.35

QUESTIONS

Find f_r, Q, and BW for this circuit. Then draw the output curve on a separate sheet of paper.

$f_r = $ _____

_____

$Q = $ _____

_____

BW $= $ _____

_____

ANSWERS

f_r = 500 Hz; Q = 31.4; BW = 16 Hz
The output curve is shown in Figure 7.36.

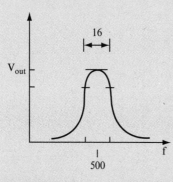

FIGURE 7.36

23 Use the circuit shown in Figure 7.37 for this problem. In this case, the resistor value is 10 ohms. However, the inductor and capacitor values are not given.

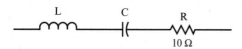

FIGURE 7.37

QUESTIONS

Find BW and the values of L and C required to give the circuit a resonant frequency of 1200 Hz and a Q of 80.

A. BW = _____

_____

B. L = _____

_____

C. C = _____

_____

ANSWERS

A. BW = 15 Hz

B. L = 106 mH

C. C = 0.166 μF

ANSWERS

A. BW = 15 Hz

B. L = 106 mH

C. C = 0.166 μF

You can check these values by using the values of L and C to find f_r.

24 Use the circuit shown in Figure 7.37 for this problem. In this case, the resistor value is given as 10 ohms. However, the inductor and capacitor values are not given.

QUESTIONS

Calculate the values of Q, L, and C required to give the circuit a resonant frequency of 300 kHz with a bandwidth of 80 kHz.

A. Q = _____

_____

B. L = _____

_____

C. C = _____

_____

ANSWERS

A. Q = 3.75

B. L = 20 μH

C. C = 0.014 μF

25 A circuit that passes (or blocks) only a narrow range of frequencies is called a *high Q* circuit. Figure 7.38 shows the output curve for a high Q circuit.

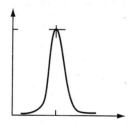

FIGURE 7.38

Because of the narrow range of frequencies it passes, a high Q circuit is said to be *selective* in the frequencies it passes.

A circuit that passes (or blocks) a wide range of frequencies is called a *low Q circuit*. Figure 7.39 shows the output curve for a low Q circuit.

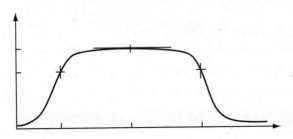

FIGURE 7.39

Recall the discussion in problem 18 (comparing the bandwidths of radio tuners and television amplifiers) to help you answer the following questions.

QUESTIONS

A. Which is the more selective, the radio tuner or the television amplifier? _____

_____

B. Which would require a lower Q circuit, the radio tuner or the television amplifier?

_____

_____

ANSWERS

A. The radio tuner

B. The television amplifier

26 The inductor and capacitor shown in Figure 7.40 are connected in parallel, rather than in series. However, you can use the same formulas you used for the series circuit in problem 20 to calculate f_r, Q, and BW for parallel LC circuits.

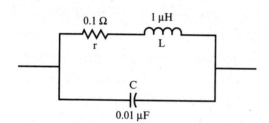

FIGURE 7.40

QUESTIONS

Find f_r, Q, and BW for the circuit shown in Figure 7.40.

A. $f_r =$ _____

B. Q = _____

C. BW = _____

ANSWERS

A. $f_r = 1.6$ MHz

B. $X_L = 10$ ohms, so Q = 10/0.1 = 100 (The only resistance here is the small DC resistance of the inductor.)

C. BW = 16 kHz (This is a fairly high Q circuit.)

27 In the last few problems, you learned how to calculate f_r, BW, and Q, for a given circuit, or conversely, to calculate the component values that would produce a circuit with specified f_r, BW, and Q values.

When you know the resonant frequency and bandwidth for a circuit, you can sketch an approximate output curve. With the simple calculations listed in this problem, you can draw a curve that is accurate to within 1 percent of its true value.

The curve that results from the calculations used in this problem is sometimes called the *general resonance curve*.

You can determine the output voltage at several frequencies by following these steps:

1. Assume the peak output voltage V_p at the resonant frequency f_r to be 100 percent. This is point A on the curve shown in Figure 7.41.

2. The output voltage at f_1 and f_2 is 0.707 of 100 percent. On the graph, these are the two points labeled B in Figure 7.41. Note that $f_2 - f_1 = $ BW. Therefore, at half a bandwidth above and below f_r, the output is 70.7 percent of V_p.

3. At f_3 and f_4 (the two points labeled C in Figure 7.41), the output voltage is 44.7 percent of V_p. Note that $f_4 - f_3 = 2$ BW. Therefore, at 1 bandwidths above and below f_r, the output is 44.7 percent of maximum.

4. At f_5 and f_6 (the two points labeled D in Figure 7.41), the output voltage is 32 percent of V_p. Note that $f_6 - f_5 = 3$ BW. Therefore, at 1.5 bandwidths above and below f_r, the output is 32 percent of maximum.

5. At f_7 and f_8 (the two points labeled E in Figure 7.41), the output voltage is 24 percent of V_p. Note that $f_8 - f_7 = 4$ BW. Therefore, at 2 bandwidths above and below f_r, the output is 24 percent of maximum.

6. At f_{10} and f_9 (the two points labeled F in Figure 7.41), the output is 13 percent of V_p. Note that $f_{10} - f_9 = 8$ BW. Therefore, at 4 bandwidths above and below f_r, the output is 13 percent of maximum.

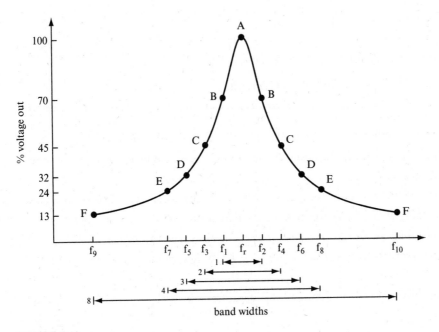

FIGURE 7.41

QUESTIONS

Calculate f_r, X_L, Q, and BW for the circuit shown in Figure 7.42.

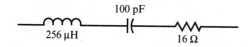

100 pF

256 µH   16 Ω

FIGURE 7.42

A. f_r = _____

_____

B. X_L = _____

_____

C. Q = _____

_____

D. BW = _____

_____

ANSWERS

A. f_r = 1 MHz

B. X_L = 1607 ohms

C. Q = 100

D. BW = 10 kHz

28 Now, calculate the frequencies that correspond with each percentage of the peak output voltage listed in steps 1 through 6 of problem 27. (Refer to the graph in Figure 7.41 as needed.)

QUESTIONS

A. At what frequency will the output level be maximum? _____

_____

B. At what frequencies will the output level be 70 percent of V_p? _____

_____

C. At what frequencies will the output level be 45 percent of V_p? _____

_____

D. At what frequencies will the output level be 32 percent of V_p? _____

_____

E. At what frequencies will the output level be 24 percent of V_p? _____

_____

F. At what frequencies will the output level be 13 percent of V_p? _____

_____

ANSWERS

A. 1 MHz

B. 995 kHz and 1005 kHz (1 MHz – 5 kHz and + 5 kHz)

C. 990 kHz and 1010 kHz

D. 985 kHz and 1015 kHz

E. 980 kHz and 1020 kHz

F. 960 kHz and 1040 kHz

29 You can calculate the output voltage at each frequency in the answers to problem 28 by multiplying the peak voltage by the related percentage for each frequency.

QUESTIONS

Calculate the output voltage for the frequencies given here, assuming that the peak output voltage is 5 volts.

A. What is the output voltage level at 995 kHz? _____

_____

B. What is the output voltage level at 980 kHz? _____

_____

ANSWERS

A. V = 5 volts × 0.70 = 3.5 volts

B. V = 5 volts × 0.24 = 1.2 volts

Figure 7.43 shows the output curve generated by plotting the frequencies calculated in problem 28 and the corresponding output voltages calculated in this problem.

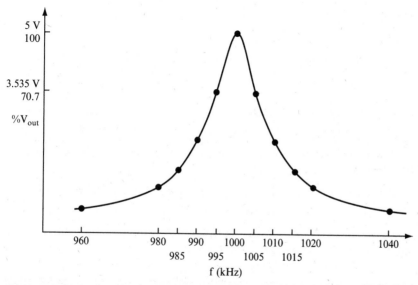

FIGURE 7.43

INTRODUCTION TO OSCILLATORS

In addition to their use in circuits used to filter input signals, capacitors and inductors are used in circuits called *oscillators*.

Oscillators are circuits that generate waveforms at particular frequencies. Many oscillators use a tuned parallel LC circuit to produce a sine wave output. This section is an introduction to the use of parallel capacitors and inductors in oscillators.

30 When the switch in the circuit shown in drawing (1) of Figure 7.44 is closed, current flows through both sides of the parallel LC circuit in the direction shown.

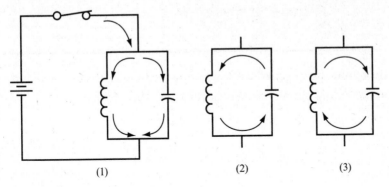

(1)   (2)   (3)

FIGURE 7.44

It is difficult for the current to flow through the inductor initially because the inductor opposes any changes in current flow. Conversely, it is easy for the current to flow into the capacitor initially because with no charge on the plates of the capacitor there is no opposition to the flow.

As the charge on the capacitor increases, the current flow in the capacitor side of the circuit decreases. However, more current flows through the inductor. Eventually, the capacitor is fully charged, so current stops flowing in the capacitor side of the circuit, and a steady current flows through the inductor.

QUESTION

When you open the switch, what happens to the charge on the capacitor? _____

_____

ANSWER

It discharges through the inductor. (Note the current direction, shown in drawing [2] of Figure 7.44.)

31 With the switch open, current continues to flow until the capacitor is fully discharged.

QUESTION

When the capacitor is fully discharged, how much current is flowing through the inductor? _____

_____

ANSWER

None.

32 Because there is no current in the inductor, its magnetic field collapses. The collapsing of the magnetic field induces a current to flow in the inductor, and this current flows in the same direction as the original current through the inductor (remember that an inductor resists any change in current flow), which is shown in drawing (2) of Figure 7.44. This current now charges the capacitor to a polarity that is opposite from the polarity that the battery induced.

QUESTION

When the magnetic field of the inductor has fully collapsed, how much current will be flowing? _____

_____

ANSWER

None.

33 Next, the capacitor discharges through the inductor again, but this time the current flows in the opposite direction, as shown in drawing (3) of Figure 7.44. The change in current direction builds a magnetic field of the opposite polarity. The magnetic field stops growing when the capacitor is fully discharged.

Because there is no current flowing through the inductor, its magnetic field collapses and induces current to flow in the direction shown in drawing (3) of Figure 7.44.

QUESTION

What do you think the current generated by the magnetic field in the inductor will do to the capacitor? _____

_____

ANSWER

It charges it to the original polarity.

34 When the field has fully collapsed, the capacitor stops charging. It now begins to discharge again, causing current to flow through the inductor in the direction shown in drawing (2) of Figure 7.44. This "seesaw" action of current will continue indefinitely.

As the current flows through the inductor, a voltage drop occurs across the inductor. The magnitude of this voltage drop will increase and decrease as the magnitude of the current changes.

QUESTION

What would you expect the voltage across the inductor to look like when you view it on an oscilloscope? _____

_____

ANSWER

A sine wave

35 In a perfect circuit, this oscillation continues and produces a continuous sine wave. In practice, a small amount of power is lost in the DC resistance of the inductor and the other wiring. As a result, the sine wave gradually decreases in amplitude and dies out to nothing after a few cycles, as shown in Figure 7.45.

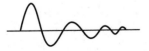

FIGURE 7.45

QUESTION

How might you prevent this fade-out? _____

_____

ANSWER

By replacing a small amount of energy in each cycle.

This lost energy can be injected into the circuit by momentarily closing and opening the switch at the correct time. (See drawing [1] of Figure 7.44.) This would sustain the oscillations indefinitely.

An electronic switch (such as a transistor) could be connected to the inductor as shown in Figure 7.46. Changes in the voltage drop across the inductor would turn the electronic switch on or off, thereby opening or closing the switch.

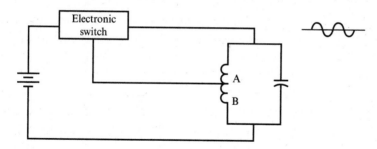

FIGURE 7.46

The small voltage drop across the few turns of the inductor (also referred to as a *coil*), between point B at the end of the coil, and point A about halfway along the coil, is used to operate the electronic switch. These points are shown in Figure 7.46.

Using a small part of an output voltage in this way is called *feedback* because the voltage is "fed back" to an earlier part of the circuit to make it operate correctly.

When you properly set up such a circuit, it produces a continuous sine wave output of constant amplitude and constant frequency. This circuit is called an *oscillator*. You can calculate the frequency of the sine waves generated by an oscillator with the following formula for determining resonant frequency:

$$f = \frac{1}{2\pi\sqrt{LC}}$$

The principles you learned in the last few problems are used in practical oscillator circuits, such as those presented in Chapter 9.

SUMMARY

In this chapter, you learned about the following topics related to resonant circuits:

- How the impedance of a series LC circuit and a parallel LC circuit changes with changes in frequency.

- At resonant frequency for a parallel LC circuit, the impedance is at its highest; whereas for a series LC circuit, impedance is at its lowest.

- The concept of bandwidth enables you to easily calculate the output voltage at various frequencies and draw an accurate output curve.

- The principles of bandpass filters and notch (or band-reject) filters.

- The fundamental concepts integral to understanding how an oscillator functions.

SELF-TEST

These questions test your understanding of the concepts covered in this chapter. Use a separate sheet of paper for your drawings or calculations. Compare your answers with the answers provided following the test.

1. What is the formula for the impedance of a series LC circuit? _____
_____

2. What is the formula for the impedance of a series RLC circuit (a circuit containing resistance, inductance, and capacitance)? _____
_____

3. What is the relationship between X_C and X_L at the resonant frequency? _____
_____

4. What is the voltage across the resistor in a series RLC circuit at the resonant frequency? _____
_____

5. What is the voltage across a resistor in series with a parallel LC circuit at the resonant frequency? _____

_____

6. What is the impedance of a series circuit at resonance? _____

_____

7. What is the formula for the impedance of a parallel circuit at resonance? _____

_____

8. What is the formula for the resonant frequency of a circuit? _____

_____

9. What is the formula for the bandwidth of a circuit? _____

_____

10. What is the formula for the Q of a circuit? _____

_____

Questions 11–13 use a series LC circuit. In each case, the values of the L, C, and R are given. Find f_r, X_L, X_C, Z, Q, and BW. Draw an output curve for each answer.

11. L = 0.1 mH, C = 0.01 μF, R = 10 ohms _____

_____

12. L = 4 mH, C = 6.4 μF, R = 0.25 ohms _____

_____

13. L = 16 mH, C = 10 μF, R = 20 ohms _____

_____

Questions 14 and 15 use a parallel LC circuit. No R is used; r is given. Find f_r, X_L, X_C, Z, Q, and BW.

14. L = 6.4 mH, C = 10 μF, r = 8 ohms _____

_____

15. L = 0.7 mH, C = 0.04 μF, r = 1.3 ohms _____

_____

16. Use the output curve shown in Figure 7.47 to answer the following questions.

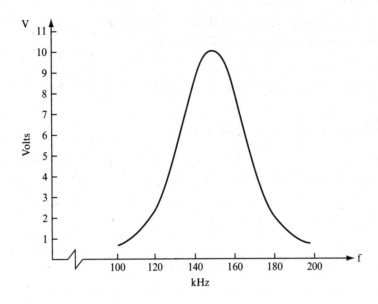

FIGURE 7.47

A. What is the peak value of the output curve? _____

_____

B. What is the resonant frequency? _____

_____

C. What is the voltage level at the half power points? _____

_____

D. What are the half power frequencies? _____

_____

E. What is the bandwidth? _____

_____

F. What is the Q of the circuit? _____

_____

ANSWERS TO SELF-TEST

If your answers do not agree with those given here, review the problems indicated in parentheses before you go on to Chapter 8, "Transistor Amplifiers."

1.	$Z = X_L - X_C$	(problem 2)
2.	$Z = \sqrt{(X_L - X_C)^2 + R^2}$	(problem 2)
3.	$X_L = X_C$	(problem 5)
4.	Maximum output	(problem 5)
5.	Minimum output	(problem 11)
6.	Z = minimum. Ideally, it is equal to the resistance.	(problem 5)
7.	$Z = \dfrac{L}{Cr}$	(problem 10)
	In this formula, r is the resistance of the coil.	
8.	$f_r = \dfrac{1}{2\pi\sqrt{LC}}$	(problem 6)
9.	$BW = \dfrac{f_r}{Q}$	(problem 20)
10.	$Q = \dfrac{X_L}{R}$	(problem 20)
	or	
	$\dfrac{X_L}{r}$	
	To draw the output curves for Questions 11–13, use the graph in Figure 7.41 as a guide and insert the appropriate bandwidth and frequency values.	(problems 21–29)
11.	f_r = 160 kHz, $X_L = X_C$ = 100 ohms, Q = 10, BW = 16 kHz, Z = 10 ohms	(problems 21–29)
12.	f_r = 1 kHz, $X_L = X_C$ = 25 ohms, Q = 100, BW = 10 Hz, Z = 0.25 ohms	(problems 21–29)

Continued

(continued)

13.	f_r = 400 Hz, X_L = X_C = 40 ohms, Q = 2, BW = 200 Hz, Z = 20 ohms	(problems 21–29)
14.	f_r = 600 Hz, X_L = 24 ohms, X_C = 26.5, Q = 3, BW = 200 Hz, Z = 80 ohms	(problems 21–29)
	Because Q is not given, you should use the more complicated of the two formulas shown in problem 10 to calculate the resonant frequency.	
15.	f_r = 30 kHz, X_L = 132 ohms, X_C = 132, Q = 101.5, BW = 300 Hz, Z = 13.4 ohms	(problems 21–29)
16A.	10.1 volts	(problems 27 and 28)
16B.	148 kHz	(problems 27 and 28)
16C.	10.1 × 0.707 = 7.14 volts	(problems 27 and 28)
16D.	Approximately 135 kHz and 160 kHz (not quite symmetrical)	(problems 27 and 28)
16E.	BW = 25 kHz	(problems 27 and 28)
16F.	$Q = \dfrac{f_r}{BW} =$ about 5.9	(problems 27 and 28)

8

Transistor Amplifiers

Many of the AC signals you'll work with in electronics are small. For example, the signal that an optical detector reads from a DVD disk cannot drive a speaker, and the signal from a microphone's output is too weak to send out as a radio signal. In cases such as these, you must use an amplifier to boost the signal.

The best way to demonstrate the basics of amplifying a weak signal to a usable level is by starting with a one-transistor amplifier. When you understand a one-transistor amplifier, you can grasp the building block that makes up amplifier circuits used in electronic devices such as cellphones, MP3 players, and home entertainment centers.

Many amplifier circuit configurations are possible. The simplest and most basic of amplifying circuits are used in this chapter to demonstrate how a transistor amplifies a signal. You can also see the steps to design an amplifier.

The emphasis in this chapter is on the bipolar junction transistor (BJT), just as it was in Chapter 3, "Introduction to the Transistor," and Chapter 4, "The Transistor Switch," which dealt primarily with the application of transistors in switching circuits. Two other types of devices used as amplifiers are also examined: the junction field effect transistor (JFET) (introduced in Chapters 3 and 4), and an integrated circuit called the *operational amplifier (op-amp)*.

When you complete this chapter, you will be able to do the following:

- Calculate the voltage gain for an amplifier.

- Calculate the DC output voltage for an amplifier circuit.

- Select the appropriate resistor values to provide the required gain to an amplifier circuit.

- Identify several ways to increase the gain of a one-transistor amplifier.

- Distinguish between the effects of a standard one-transistor amplifier and an emitter follower circuit.

- Design a simple emitter follower circuit.

- Analyze a simple circuit to find the DC level out and the AC gain.

- Design a simple common source (JFET) amplifier.

- Analyze a JFET amplifier to find the AC gain.

- Recognize an op-amp and its connections.

WORKING WITH TRANSISTOR AMPLIFIERS

1 In Chapter 3 you learned how to turn transistors ON and OFF. You also learned how to calculate the value of resistors in amplifier circuits to set the collector DC voltage to half the power supply voltage. To review this concept, examine the circuit shown in Figure 8.1.

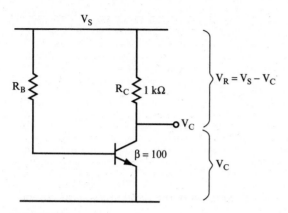

FIGURE 8.1

Use the following steps to find the value of R_B that will set the collector DC voltage (V_C) to half the supply voltage (V_S):

1. Find I_C by using the following equation:

$$I_C = \frac{V_R}{R_C} = \frac{V_S - V_C}{R_C}$$

2. Find I_B by using the following equation:

$$I_B = \frac{I_C}{\beta}$$

3. Find R_B by using the following equation:

$$R_B = \frac{V_S}{I_B}$$

QUESTIONS

Find the value of R_B that will set the collector voltage to 5 volts, using steps 1–3 and the following values for the circuit:

$$V_S = 10 \text{ volts}, \quad R_C = 1 \text{k}\Omega, \quad \beta = 100$$

A. $I_C = $ _____

B. $I_B = $ _____

C. $R_B = $ _____

ANSWERS

A. $I_C = \dfrac{5 \text{ volts}}{1 \text{k}\Omega} = 5 \text{mA}$

B. $I_B = \dfrac{5}{100} = 0.05 \text{mA}$

C. $R_B = \dfrac{10 \text{ volts}}{0.05 \text{mA}} = 200 \text{k}\Omega$

2 You have seen that using a 200 kΩ resistor for R_B gives an output level of 5 volts at the collector. This procedure of setting the output DC level is called *biasing*. In problem 1, you biased the transistor to a 5-volt DC output.

Use the circuit shown in Figure 8.1 and the formulas given in problem 1 to answer the following questions.

QUESTIONS

A. If you decrease the value of R_B, how do I_B, I_C, V_R, and the bias point V_C change?____

_____

B. If you increase the value of R_B, how do I_B, I_C, V_R, and V_C change? _____

_____

ANSWERS

A. I_B increases, I_C increases, V_R increases, and so the bias point V_C decreases.

B. I_B decreases, I_C decreases, V_R decreases, and so the bias point V_C increases.

3 In problem 2, you found that changing the value of R_B in the circuit shown in Figure 8.1 changes the value of I_B.

The transistor amplifies slight variations in I_B. Therefore, the amount I_C fluctuates is β times the change in value in I_B.

The variations in I_C cause changes in the voltage drop V_R across R_C. Therefore, the output voltage measured at the collector also changes.

QUESTIONS

For the circuit shown in Figure 8.1, calculate the following parameters when $R_B = 168$ kΩ and $V_S = 10$ volts:

A. $I_B = \dfrac{V_C}{R_B} = $ _____

B. $I_C = \beta I_B = $ _____

C. $V_R = I_C R_C = $ _____

D. $V_C = V_S - V_R = $ _____

ANSWERS

A. $I_B = \dfrac{10\,\text{volts}}{168\ \text{k}\Omega} = 0.059\,\text{mA}$

B. $I_C = 100 \times 0.059 = 5.9\ \text{mA}$

C. $V_R = 1\ \text{k}\Omega \times 5.9\ \text{mA} = 5.9\ \text{volts}$

D. $V_C = 10\ \text{volts} - 5.9\ \text{volts} = 4.1\ \text{volts}$

4 Use the circuit shown in Figure 8.1 to answer the following questions when $V_S = 10$ volts.

QUESTIONS

Calculate V_C for each of the following values of R_B:

A. 100 kΩ _____

B. 10 MΩ _____

C. 133 kΩ _____

D. 400 kΩ _____

ANSWERS

A. $I_B = 0.1\ \text{mA}$, $I_C = 10\ \text{mA}$, $V_C = 0$ volts

B. $I_B = 1\ \mu\text{A}$, $I_C = 0.1\ \text{mA}$, $V_C = 10$ volts (approximately)

(Continued)

(continued)

5 The values of I_C and V_C that you calculated in problems 1 and 4 are plotted on the graph on the left side of Figure 8.2. The straight line connecting these points on the graph is called the *load line.*

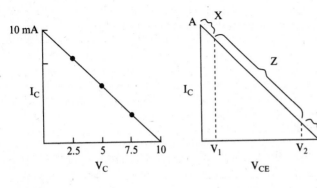

FIGURE 8.2

The axis labeled V_C represents the voltage between the collector and the emitter of the transistor, and not the voltage between the collector and ground. Therefore, this axis should correctly be labeled V_{CE}, as shown in the graph on the right of the figure. (For this circuit, $V_{CE} = V_C$ because there is no resistor between the emitter and ground.)

QUESTIONS

A. At point A in the graph on the right, is the transistor ON or OFF? _____

B. Is it ON or OFF at point B?_____

B. OFF because essentially no current flows, and the transistor acts like an open circuit. The voltage drop across the transistor is at its maximum (10 volts, in this case).

6 Point A on the graph shown in Figure 8.2 is called the *saturated point* (or the *saturation point*) because it is at that point that the collector current is at its maximum.

Point B on the graph shown in Figure 8.2 is often called the *cutoff* point because, at that point, the transistor is OFF and no collector current flows.

In regions X and Y, the gain (β) is not constant, so these are called the *nonlinear regions*. Note that $\beta = I_C/I_B$.

As a rough guide, V_1 is approximately 1 volt, and V_2 is approximately 1 volt less than the voltage at point B.

QUESTION

What is the value of V_{CE} at point B? _____

_____

ANSWER

$V_{CE} = V_S$, which is 10 volts in this case.

7 In region Z of the graph shown in Figure 8.2, β (that is, the slope of the graph) is constant (at a given temperature). Therefore, this is called the *linear region*. Operating the transistor in the linear region results in an output signal that is free of distortion.

QUESTION

Which values of I_C and V_C would result in an undistorted output in the circuit shown in Figure 8.1?

A $I_C = 9$ mA, $V_C = 1$ volt

B. $I_C = 1$ mA, $V_C = 9$ volts

C. $I_C = 6$ mA, $V_C = 4.5$ volts _____

_____

ANSWER

C is the only one. A and B fall into nonlinear regions.

8 If you apply a small AC signal to the base of the transistor after it has been biased, the small voltage variations of the AC signal (shown in Figure 8.3 as a sine wave) cause small variations in the base current.

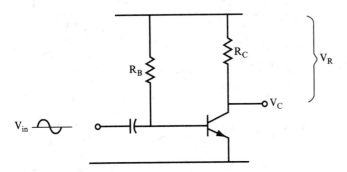

FIGURE 8.3

These variations in the base current will be amplified by a factor of β and will cause corresponding variations in the collector current. The variations in the collector current, in turn, will cause similar variations in the collector voltage.

The β used for AC gain calculations is different from the β used in calculating DC variations. The AC β is the value of the common emitter AC forward current transfer ratio, which is listed as h_{fe} in the manufacturer's data sheets for the transistor. Use the AC β whenever you need to calculate the AC output for a given AC input, or to determine an AC current variation. Use the DC β to calculate the base or collector DC current values. You must know which β to use, and remember that one is used for DC, and the other is used for AC variations. The DC β is sometimes called h_{FE} or $β_{dc}$.

As V_{in} increases, the base current increases, which causes the collector current to increase. An increase in the collector current increases the voltage drop across R_C, which causes V_C to decrease.

NOTE The capacitor shown at the input blocks DC (infinite reactance) and easily passes AC (low reactance). This is a common isolation technique used at the input and output of AC circuits.

QUESTIONS

A. If the input signal decreases, what happens to the collector voltage? _____

_____

B. If you apply a sine wave to the input, what waveform would you expect at the collector? _____

_____

ANSWERS

A. The collector voltage, V_C, increases.

B. A sine wave, but inverted as shown in Figure 8.4.

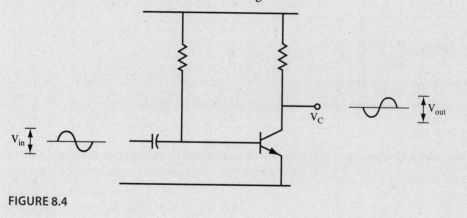

FIGURE 8.4

9 Figure 8.4 shows the input and output sine waves for an amplifier circuit.

The input voltage V_{in} is applied to the base. (Strictly speaking, it is applied across the base-emitter diode.) The voltage variations at the collector are centered on the DC bias point V_C, and they will be larger than variations in the input voltage. Therefore, the output sine wave is larger than the input sine wave (that is, amplified).

This amplified output signal at the collector can be used to drive a load (such as a speaker).

To distinguish these AC variations in output from the DC bias level, you indicate the AC output voltage by V_{out}. In most cases, V_{out} is a peak-to-peak value.

QUESTIONS

A. What is meant by V_C? _____

_____

B. What is meant by V_{out}? _____

_____

ANSWERS

A. Collector DC voltage, or the bias point

B. AC output voltage

The ratio of the output voltage to the input voltage is called the *voltage gain* of the amplifier.

$$\text{Voltage gain} = A_V = \frac{V_{out}}{V_{in}}$$

To calculate the voltage gain of an amplifier, you can measure V_{in} and V_{out} with an oscilloscope. Measure peak-to-peak voltages for this calculation.

10 For the circuit shown in Figure 8.4 you can calculate the voltage gain using the following formula:

$$A_V = \beta \times \frac{R_L}{R_{in}}$$

In this equation:

- R_L is the *load resistance*. In this circuit, the collector resistor, R_C, is the load resistance.

- R_{in} is the *input resistance* of the transistor. You can find R_{in} (often called h_{ie}) on the data or specification sheets from the manufacturer. In most transistors, input resistance is approximately 1 kΩ to 2 kΩ.

You can find V_{out} by combining these two voltage gain equations:

$$A_V = \frac{V_{out}}{V_{in}} \quad \text{and} \quad A_V = \beta \times \frac{R_L}{R_{in}}$$

$$\text{Therefore,} \quad \frac{V_{out}}{V_{in}} = \beta \times \frac{R_L}{R_{in}}$$

Solving this for V_{out} results in the following equation. Here, the values of $R_{in} = 1$ kΩ, $V_{in} = 1$ mV, $R_C = 1$ kΩ, and $β = 100$ were used to perform this sample calculation.

$$V_{out} = V_{in} \times β \times \frac{R_L}{R_{in}}$$

$$= 1 \text{mV} \times 100 \times \frac{1 \text{k}Ω}{1 \text{k}Ω}$$

$$= 100 \text{mV}$$

QUESTIONS

A. Calculate V_{out} if $R_{in} = 2$ kΩ, $V_{in} = 1$ mV, $R_C = 1$ kΩ, and $β = 100$. _____

_____

B. Find the voltage gain in both cases. _____

_____

ANSWERS

A. $V_{out} = 50$ mV

B. $A_V = 100$ and $A_V = 50$

This simple amplifier can provide voltage gains of up to approximately 500. But it does have several faults that limit its practical usefulness.

- Because of variations in $β$ between transistors, V_C changes if the transistor is changed. To compensate for this, you must adjust R_B.

- R_{in} or h_{ie} varies greatly from transistor to transistor. This variation, combined with variations in $β$, means that you cannot guarantee the gain from one transistor amplifier to another.

- Both R_{in} and $β$ change greatly with temperature; hence the gain is temperature-dependent. For example, a simple amplifier circuit like that discussed in this problem was designed to work in the desert in July. It would fail completely in Alaska in the winter. If the amplifier worked perfectly in the lab, it probably would not work outdoors on either a hot or cold day.

NOTE   An amplifier whose gain and DC level bias point change as described in this problem is said to be *unstable*. For reliable operation, an amplifier should be as stable as possible. In later problems, you see how to design a stable amplifier.

A STABLE AMPLIFIER

11 You can overcome the instability of the transistor amplifier discussed in the first ten problems of this chapter by adding two resistors to the circuit. Figure 8.5 shows an amplifier circuit to which resistors R_E and R_2 have been added. R_2, along with R_1 (labeled R_B in the previous circuits), ensures the stability of the DC bias point.

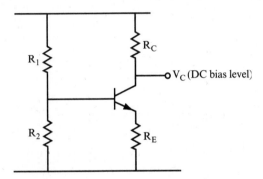

FIGURE 8.5

By adding the emitter resistor R_E, you ensure the stability of the AC gain.

The labels in Figure 8.6 identify the DC currents and voltages present in the circuit. These parameters are used in the next several problems.

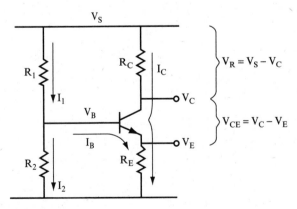

FIGURE 8.6

QUESTION

In designing an amplifier circuit and choosing the resistor values, there are two goals. What are they? _____

_____

ANSWER

A stable DC bias point, and a stable AC gain

12 Look at the gain first. The gain formula for the circuit shown in Figure 8.6 is as follows:

$$A_V = \frac{V_{out}}{V_{in}} = \frac{R_C}{R_E}$$

This is a slight variation on the formula shown in problem 10. (The complex mathematical justification for this is not important right here.) Here, the AC gain is not affected by transistor β and transistor input resistance, so the AC gain will be constant, regardless of variations in these parameters.

QUESTIONS

Use the circuit shown in Figure 8.6 with $R_C = 10$ kΩ and $R_E = 1$ kΩ to answer the following questions:

A. What is the AC voltage gain for a transistor if its $\beta = 100$? _____

_____

B. What is the gain if $\beta = 500$? _____

_____

ANSWERS

A. 10

B. 10

13 This problem provides a couple of examples that can help you understand how to calculate voltage gain and the resulting output voltage.

QUESTIONS

A. Calculate the voltage gain (A_V) of the amplifier circuit shown in Figure 8.6 if $R_C = 10\ k\Omega$ and $R_E = 1\ k\Omega$. Then, use A_V to calculate the output voltage if the input signal is 2 mV_{pp}. _____

_____

B. Calculate the voltage gain if $R_C = 1\ k\Omega$ and $R_E = 250$ ohms. Then, use A_V to calculate the output voltage if the input signal is 1 V_{pp}. _____

_____

ANSWERS

A. $A_V = \dfrac{R_C}{R_E} = \dfrac{10\,k\Omega}{1\,k\Omega} = 10$

$V_{out} = 10 = V_{in} = 20\,mV$

B. $A_V = \dfrac{1\,k\Omega}{250\,ohms} = 4$

$V_{out} = 4\,V_{pp}$

Although the amplifier circuit shown in Figure 8.6 produces stable values of voltage gain, it does not produce high values of voltage gain. For various reasons, this circuit is limited to voltage gains of 50 or less. Later, this chapter discusses an amplifier circuit that can produce higher values of voltage gain.

14 Before you continue, look at the current relationships in the amplifier circuit shown in Figure 8.6 and an approximation that is often made. You can calculate the current flowing through the emitter resistor with the following equation:

$I_E = I_B + I_C$

In other words, the emitter current is the sum of the base and the collector currents. I_C is much larger than I_B. You can, therefore, assume that the emitter current is equal to the collector current.

$I_E = I_C$

QUESTIONS

Calculate V_C, V_E, and A_V for the circuit shown in Figure 8.7 with $V_S = 10$ volts, $I_C = 1$ mA, $R_C = 1$ kΩ, and $R_E = 100$ ohms. _____

_____

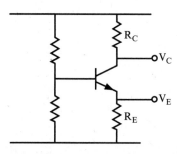

FIGURE 8.7

ANSWER

$V_R = 1\text{k}\Omega \times 1\text{mA} = 1$ volt

$V_C = V_S - V_R = 10 - 1 = 9$ volts

$V_E = 100 \text{ ohms} \times 1\text{mA} = 0.1$ volt

$A_V = \dfrac{R_C}{R_E} = \dfrac{1\text{k}\Omega}{100 \text{ ohms}} = 10$

15 For this problem, use the circuit shown in Figure 8.7 with $V_S = 10$ volts, $I_C = 1$ mA, $R_C = 2$ kΩ, and $R_E = 1$ kΩ.

QUESTION

Calculate V_C, V_E, and A_V. _____

_____

ANSWERS

$V_R = 2\,k\Omega \times 1\,mA = 2$ volts

$V_C = 10 - 2 = 8$ volts

$V_E = 1\,k\Omega \times 1\,mA = 1$ volt

$A_V = \dfrac{R_C}{R_E} = \dfrac{2\,k\Omega}{1\,k\Omega} = 2$

16 For this problem, use the circuit shown in Figure 8.7 with $V_s = 10$ volts and $I_C = 1$ mA.

QUESTIONS

Find V_C, V_E, and A_V for the following values of R_C and R_E:

A. $R_C = 5\,k\Omega$, $R_E = 1\,k\Omega$ _____

_____

B. $R_C = 4.7\,k\Omega$, $R_E = 220$ ohms _____

_____

ANSWERS

A. $V_R = 5$ volts, $\quad V_C = 5$ volts, $\quad V_E = 1$ volt, $\quad A_V = 5$

B. $V_R = 4.7$ volts, $\quad V_C = 5.3$ volts, $\quad V_E = 0.22$ volts, $\quad A_V = 21.36$

BIASING

17 In this problem, you see the steps used to calculate the resistor values needed to bias the amplifier circuit shown in Figure 8.8.

You can determine values for R_1, R_2, and R_E that bias the circuit to a specified DC output voltage and a specified AC voltage gain by using the following steps.

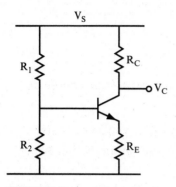

FIGURE 8.8

Read the following procedure and the relevant formulas first, and then you will work through an example.

1. Find R_E by using the following:

$$A_V = \frac{R_C}{R_E}.$$

2. Find V_E by using the following:

$$A_V = \frac{V_R}{V_E} = \frac{V_S - V_C}{V_E}.$$

3. Find V_B by using the following:

$$V_B = V_E + 0.7 \text{ volt}$$

4. Find I_C by using the following:

$$I_C = \frac{V_S - V_C}{R_C}.$$

5. Find I_B by using the following:

$$I_B = \frac{I_C}{\beta}$$

6. Find I_2 where I_2 is $10I_B$. (Refer to the circuit shown in Figure 8.6.) This is a convenient rule of thumb that is a crucial step in providing stability to the DC bias point.

7. Find R_2 by using the following:

$$R_2 = \frac{V_B}{I_2}.$$

8. Find R_1 by using the following:
$$R_1 = \frac{V_S - V_B}{I_2 + I_B}.$$

9. Steps 7 and 8 might produce nonstandard values for the resistors, so choose the nearest standard values.

10. Use the voltage divider formula to see if the standard values you chose in step 9 result in a voltage level close to V_B found in step 3. ("Close" means within 10 percent of the ideal.)

This procedure produces an amplifier that works, and results in a DC output voltage and AC gain that are close to those specified at the beginning of the problem.

QUESTIONS

Find the values of the parameters specified in each of the following questions for the circuit shown in Figure 8.9 if $A_V = 10$, $V_C = 5$ volts, $R_C = 1$ kΩ, $\beta = 100$, and $V_S = 10$ volts.

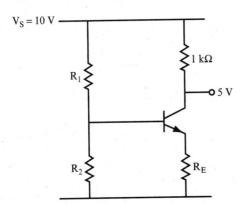

FIGURE 8.9

Work through steps 1–10, referring to the steps in this problem for formulas as necessary.

1. Find R_E.
$$A_V = \frac{R_C}{R_E}. \text{ So } R_E = \frac{R_C}{A_V} = \frac{1\text{k}\Omega}{10} = 100\,\text{ohms.}$$

2. $V_E = $ _____

3. $V_B = $ _____

4. $I_C =$ _____

5. $I_B =$ _____

6. $I_2 =$ _____

7. $R_2 =$ _____

8. $R_1 =$ _____

9. Choose the standard resistance values that are closest to the calculated values for R_1 and R_2.

   $R_1 =$ _____

   $R_2 =$ _____

10. Using the standard resistance values for R_1 and R_2, find V_B.

   $V_B =$ _____

ANSWERS

You should have found values close to the following:

1. 100 ohms

2. 0.5 volt

3. 1.2 volts

4. 5 mA

5. 0.05 mA

6. 0.5 mA

7. 2.4 kΩ

8. 16 kΩ

9. 2.4 kΩ and 16 kΩ are standard values. (They are 5 percent values.) Alternative acceptable values would be 2.2 kΩ and 15 kΩ.

10. With 2.4 kΩ and 16 kΩ, V_B = 1.3 volts. With 2.2 kΩ and 15 kΩ, V_B = 1.28 volts. Either value of V_B is within 10 percent of the 1.2 volts calculated for V_B in step 3.

Figure 8.10 shows an amplifier circuit using the values you calculated in this problem for R_1, R_2, and R_E.

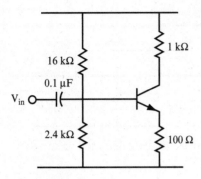

FIGURE 8.10

18 Follow the steps in problem 17 to answer the following questions.

QUESTIONS

Find the values of the parameters specified in each question for the circuit shown in Figure 8.9 if $A_V = 15$, $V_C = 6$ volts, $\beta = 100$, $R_C = 3.3$ kΩ, and $V_S = 10$ volts.

1. $R_E =$ _____

2. $V_E =$ _____

3. $V_B =$ _____

4. $I_C =$ _____

5. $I_B =$ _____

6. $I_2 =$ _____

7. $R_2 =$ _____

8. $R_1 =$ _____

9. Choose the standard resistance values that are closest to the calculated values for R_1 and R_2.

   $R_1 =$ _____

   $R_2 =$ _____

10. Using the standard resistance values for R_1 and R_2, find V_B.

   $V_B =$ _____

ANSWERS

Following are the values you should have found:

1. 220 ohms

2. 0.27 volt

3. 0.97 volt (You can use 1 volt if you want.)

4. 1.2 mA

5. 0.012 mA

6. 0.12 mA

7. 8.3 kΩ

8. 68.2 kΩ

9. These are close to the standard values of 8.2 kΩ and 68 kΩ.

10. 1.08 volts using the standard values. This is close enough to the value of V_B calculated in question 3.

PROJECT 8.1: The Transistor Amplifier

OBJECTIVE

The objective of this project is to demonstrate how AC voltage gain changes when you use resistors of different values and transistors with different current gain in a transistor amplifier circuit.

GENERAL INSTRUCTIONS

When the circuit is set up, you measure V_{out} for each set of resistors, and find A_V, using the ratio V_{out}/V_{in}. You also determine a calculated A_V using the ratio R_C/R_E in each case to determine how close the calculated A_V is to the measured A_V. You repeat this measurement with a second transistor for each set of resistors.

Parts List

You need the following equipment and supplies:

❑ One 1 kΩ, 0.25-watt resistor

❑ One 100 Ω, 0.25-watt resistor

❑ One 15 kΩ, 0.25-watt resistor

❑ One 2.2 kΩ, 0.25-watt resistor

❑ One 3.3 kΩ, 0.25-watt resistor

❑ One 220 Ω, 0.25-watt resistor

❑ One 68 kΩ, 0.25-watt resistor

❑ One 8.2 kΩ, 0.25-watt resistor

❑ One 0.1 µF capacitor

❑ One lab type power supply or 9-volt battery

❑ One function generator.

❑ One oscilloscope

❑ One breadboard

❑ One 2N3904 transistor

❑ One PN2222 transistor

Figure 8.11 shows the pinout diagram for 2N3904 and PN2222 transistors.

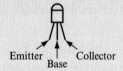

Emitter   Collector
      Base

FIGURE 8.11

STEP-BY-STEP INSTRUCTIONS

Set up the circuit shown in Figure 8.12 using the components listed for Circuit # 1 in the following table. If you have some experience in building circuits, this schematic (along with the previous parts list) should provide all the information you need to build the circuit. If you need a bit more help building the circuit, look at the photos of the completed circuit in the "Expected Results" section. (If you don't have a lab type power supply to provide 10 volts as indicated on the schematic, use a 9-volt battery.)

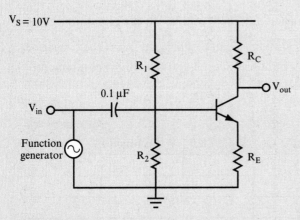

FIGURE 8.12

Circuit #	Transistor	R_C	R_E	R_1	R_2
1	PN2222	1 kΩ	100 Ω	15 kΩ	2.2 kΩ
2	2N3904	1 kΩ	100 Ω	15 kΩ	2.2 kΩ
3	PN2222	3.3 kΩ	220 Ω	68 kΩ	8.2 kΩ
4	2N3904	3.3 kΩ	220 Ω	68 kΩ	8.2 kΩ

Carefully check your circuit against the diagram.

After you check your circuit, follow these steps, and record your measurements in the blank table following the steps.

1. Connect the oscilloscope probe for channel 2 to a jumper wire connected to V_{in}, and then connect the ground clip to a jumper wire attached to the ground bus.

2. Connect the oscilloscope probe for channel 1 to a jumper wire connected to V_{out}, and then connect the ground clip to a jumper wire attached to the ground bus.

3. Set the function generator to generate a 10 kHz sine wave with approximately 0.2 V_{pp}.

4. Measure and record V_{in} and V_{out}.

5. Change the components to those listed in the next row of the table (Circuit # 2 in this case.) You should turn off the power to the circuit before changing components to avoid shorting leads together.

6. Measure and record V_{in} and V_{out}.

7. Repeat steps 5 and 6 until you have recorded V_{in} and V_{out} in the last row of the table.

8. Determine β for each of the transistors used in this project. Insert the transistors one at a time into the circuit you built in Project 3-1 to take this measurement.

9. For each transistor, record β in the following table.

Circuit #	Calculated A_V (R_C/R_E)	Transistor	β	V_{in} (Volts)	V_{out} (Volts)	Measured A_V (V_{in}/V_{out})
1	10	PN2222				
2	10	2N3904				
3	15	PN2222				
4	15	2N3904				

EXPECTED RESULTS

Figure 8.13 shows the breadboarded Circuit # 1.

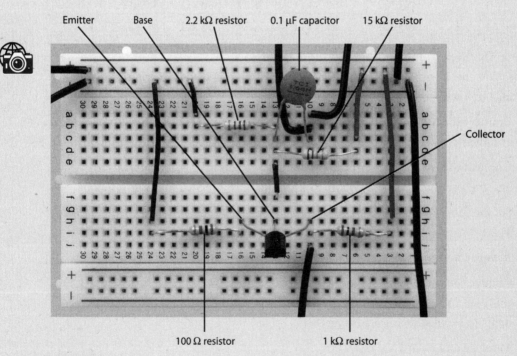

Emitter   Base   2.2 kΩ resistor   0.1 µF capacitor   15 kΩ resistor

Collector

100 Ω resistor   1 kΩ resistor

FIGURE 8.13

Figure 8.14 shows the breadboarded Circuit # 3.

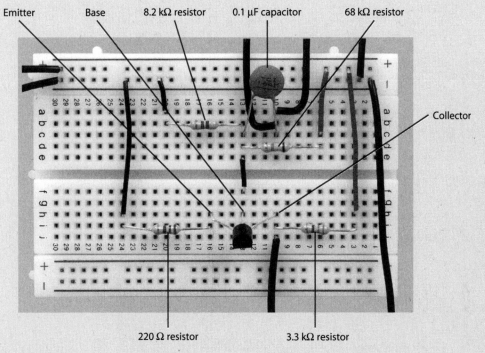

Emitter   Base   8.2 kΩ resistor   0.1 μF capacitor   68 kΩ resistor

Collector

220 Ω resistor   3.3 kΩ resistor

FIGURE 8.14

Figure 8.15 shows a function generator and oscilloscope attached to the circuit.

The input signal is represented by the upper sine wave shown in Figure 8.16, and the output signal is represented by the lower sine wave. Read the number of divisions for the peak-to-peak output sine wave, and multiply it by the corresponding VOLTS/DIV setting to determine V_{out}.

As you measure V_{in} and V_{out} for each circuit, you may need to adjust the TIME/DIV control, VOLTS/DIV control, and vertical POSITION controls on the oscilloscope. The controls shown in Figure 8.17 are adjusted to measure V_{in} and V_{out} for Circuit # 2.

Your values should be close to those shown in the following table.

Circuit #	Calculated A_V (R_C/R_E)	Transistor	β	V_{in} (Volts)	V_{out} (Volts)	Measured A_V (V_{in}/V_{out})
1	10	PN2222	235	0.22	2.1	9.5
2	10	2N3904	174	0.22	2.1	9.5
3	15	PN2222	235	0.22	3.2	14.5
4	15	2N3904	174	0.22	3.2	14.5

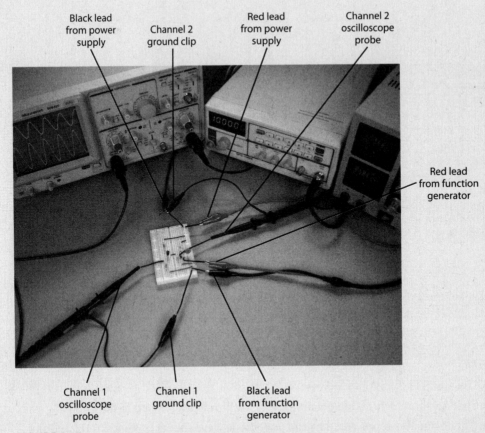

Black lead from power supply

Channel 2 ground clip

Red lead from power supply

Channel 2 oscilloscope probe

Red lead from function generator

Channel 1 oscilloscope probe

Channel 1 ground clip

Black lead from function generator

FIGURE 8.15

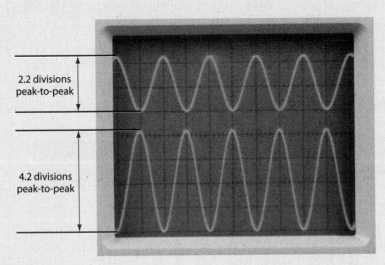

2.2 divisions peak-to-peak

4.2 divisions peak-to-peak

FIGURE 8.16

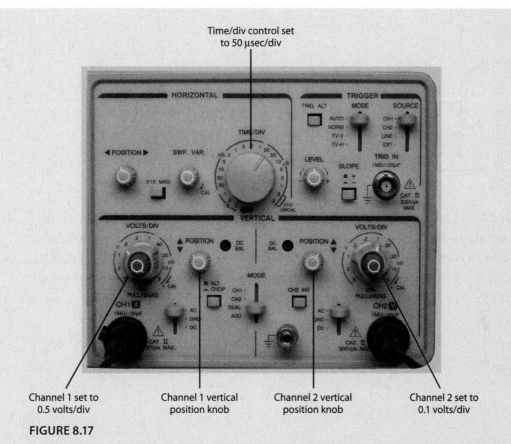

Time/div control set
to 50 μsec/div

Channel 1 set to
0.5 volts/div

Channel 1 vertical
position knob

Channel 2 vertical
position knob

Channel 2 set to
0.1 volts/div

FIGURE 8.17

The measured values of A_V are quite close to the calculated values of A_V, well within variations that could be caused by the ± 5 percent tolerance specified for the resistor values. Also, the variation in transistor β had no effect on the measured values of A_V.

19 The AC voltage gain for the circuit discussed in problem 18 was 15. Earlier, you learned that the maximum practical gain of the amplifier circuit shown in Figure 8.9 is approximately 50.

However, in problem 10, you learned that AC voltage gains of up to 500 are possible for the amplifier circuit shown in Figure 8.4. Therefore, by ensuring the stability of the DC bias point, the amplifier has much lower gain than is possible with the transistor amplifier circuit shown in Figure 8.4.

You can make an amplifier with stable bias points without giving up high AC voltage gain by placing a capacitor in parallel with the emitter resistor, as shown in Figure 8.18.

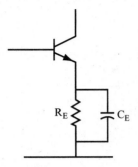

FIGURE 8.18

If the reactance of this capacitor for an AC signal is significantly smaller than R_E, the AC signal passes through the capacitor rather than the resistor. Therefore, the capacitor is called an *emitter bypass capacitor*. The AC signal "sees" a different circuit from the DC, which is blocked by the capacitor and must flow through the resistor. Figure 8.19 shows the different circuits seen by AC and DC signals.

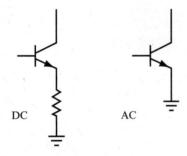

FIGURE 8.19

The AC voltage gain is now close to that of the amplifier circuit discussed in problems 1–10.

QUESTIONS

A. What effect does the emitter bypass capacitor have on an AC signal? _____
_____

B. What effect does the emitter bypass capacitor have on the AC voltage gain? _____
_____

C. What is the AC voltage gain formula with an emitter bypass capacitor included in the circuit? _____
_____

ANSWERS

A. It makes the emitter look like a ground and effectively turns the circuit into the circuit shown in Figure 8.4.

B. It increases the gain.

C. The same formula used in problem 10:

$$A_V = \beta \times \frac{R_C}{R_{in}}$$

---

20 You can use the circuit shown in Figure 8.18 when you need as much AC voltage gain as possible. When high AC voltage gain is your priority, predicting the actual amount of gain is usually not important, so the fact that the equation is inexact is unimportant. If you need an accurate amount of gain, then you must use a different type of amplifier circuit that produces lower amounts of gain.

You can find the value of the capacitor C_E using the following steps:

1. Determine the lowest frequency at which the amplifier must operate.

2. Calculate X_C with the following formula:

$$X_C = \frac{R_E}{10}$$

3. Calculate C_E with the following formula using the lowest frequency at which the amplifier must operate (determined in step 1):

$$X_C = \frac{1}{2\pi fC}$$

For the following question, use the circuit shown in Figure 8.10, with an emitter bypass capacitor added, as shown in Figure 8.20.

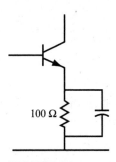

FIGURE 8.20

QUESTIONS

Follow the previous steps to calculate the value of C_E required if the lowest operating frequency of the amplifier is 50 Hz.

1. 50 Hz is the lowest frequency at which the amplifier must operate.

2. $X_C =$ _____

3. $C_E =$ _____

ANSWERS

$X_C = 10$ ohms

$C_E = 320\ \mu F$ (approximately)

The AC voltage gain formula for an amplifier with an emitter bypass capacitor (Circuit 2 in Figure 8.21) is the same as the AC voltage gain formula for the amplifiers discussed in problems 1–10, where the emitter is directly connected to ground (Circuit 1 in Figure 8.21).

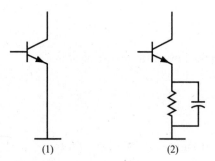

(1)   (2)

FIGURE 8.21

The AC voltage gain formula for an amplifier is as follows:

$$A_V = \beta \times \frac{R_C}{R_{in} + R_E}$$

(R_C is used instead of R_L because the collector resistor is the total load on the amplifier.)

- *Circuit 1*—Here, R_E = zero, so the AC voltage gain formula is as follows:

$$A_V = \beta \times \frac{R_C}{R_{in}}.$$

- *Circuit 2*—Here, R_E = zero for an AC signal because the AC signal is grounded by the capacitor, and R_E is out of the AC circuit. Thus, the AC voltage gain formula is as follows:

$$A_V = \beta \times \frac{R_C}{R_{in}}.$$

21 To obtain even larger voltage gains, two transistor amplifiers can be *cascaded*. That is, you can feed the output of the first amplifier into the input of the second amplifier. Figure 8.22 shows a two-transistor amplifier circuit, also called a *two-stage amplifier*.

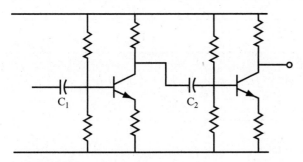

FIGURE 8.22

You find the total AC voltage gain by multiplying the individual gains. For example, if the first amplifier has an AC voltage gain of 10, and the second has an AC voltage gain of 10, then the overall AC voltage gain is 100.

QUESTIONS

A. Suppose you cascade an amplifier with a gain of 15 with one that has a gain of 25. What is the overall gain? _____

_____

B. What is the overall gain if the individual gains are 13 and 17? _____

_____

22 Two-stage amplifiers can achieve large AC voltage gains if each amplifier uses an emitter bypass capacitor.

QUESTION

What is the total AC voltage gain if each stage of a two-transistor amplifier has a gain of 100? _____

_____

ANSWER

10,000

THE EMITTER FOLLOWER

23 Figure 8.23 shows another type of amplifier circuit.

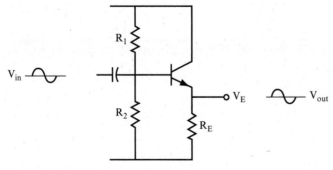

FIGURE 8.23

QUESTION

How is the circuit shown in Figure 8.23 different from the amplifier circuit discussed in problems 11–18? _____

_____

ANSWER

There is no collector resistor, and the output signal is taken from the emitter.

24 The circuit shown in Figure 8.23 is called an *emitter follower* amplifier. (In some cases, it is also called the *common collector* amplifier.)

The output signal has some interesting features:

- The peak-to-peak value of the output signal is almost the same as the input signal. In other words, the circuit gain is slightly less than 1; although in practice it is often considered to be 1.

- The output signal has the same phase as the input signal. It is not inverted; the output is simply considered to be the same as the input.

- The amplifier has a high input resistance. Therefore, it draws little current from the signal source.

- The amplifier has a low output resistance. Therefore, the signal at the emitter appears to be emanating from a battery or signal generator with a low internal resistance.

QUESTIONS

A. What is the voltage gain of an emitter follower amplifier? _____

_____

B. Is the output signal inverted? _____

_____

C. What is the input resistance of the emitter follower amplifier? _____

_____

D. What is its output resistance? _____

_____

ANSWERS

A. 1

B. No

C. High

D. Low

25 The example in this problem demonstrates the importance of the emitter follower circuit. The circuit shown in Figure 8.24 contains a small AC motor with 100 ohms resistance that is driven by a 10 V$_{pp}$ signal from a generator. The 50-ohm resistor labeled R$_G$ is the internal resistance of the generator. In this circuit, only 6.7 V$_{pp}$ is applied to the motor; the rest of the voltage is dropped across R$_G$.

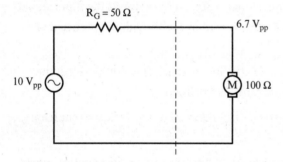

FIGURE 8.24

Figure 8.25 shows the same circuit, with a transistor connected between the generator and the motor in an emitter follower configuration.

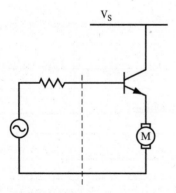

FIGURE 8.25

You can use the following formula to calculate the approximate input resistance of the transistor:

$$R_{in} = \beta \times R_E = 100 \times 100\,\Omega = 10{,}000\,\Omega \text{ (assuming that } \beta = 100)$$

The 10 V_{pp} from the generator is divided between the 10,000-ohm input resistance of the transistor and the 50-ohm internal resistance of the generator. Therefore, there is no significant voltage drop across R_G, and the full 10 V_{pp} is applied to the base of the transistor. The emitter voltage remains at 10 V_{pp}.

Also, the current through the motor is now produced by the power supply and not the generator, and the transistor looks like a generator with a low internal resistance.

This internal resistance (R_O) is called the *output impedance* of the emitter follower. You can calculate it using this formula:

$$R_O = \frac{\text{internal resistance of generator}}{\beta}$$

For the circuit shown in Figure 8.25, if R_G = 50 ohms and β = 100, R_O = 0.5 ohms. Therefore, the circuit shown in Figure 8.25 is effectively a generator with an internal resistance of only 0.5 ohms driving a motor with a resistance of 100 ohms. Therefore, the output voltage of 10 V_{pp} is maintained across the motor.

QUESTIONS

A. What is the emitter follower circuit used for in this example? _____

_____

B. Which two properties of the emitter follower are useful in circuits? _____

_____

ANSWERS

A. To drive a load that could not be driven directly by a generator

B. High input resistance and its low output resistance

26 The questions in this problem apply to the emitter follower circuit discussed in problems 23–25.

QUESTIONS

A. What is the approximate gain of an emitter follower circuit? _____

_____

B. What is the phase of the output signal compared to the phase of the input signal? _

_____

C. Which has the higher value, the input resistance or the output resistance? _____

_____

D. Is the emitter follower more effective at amplifying signals or at isolating loads? ___

_____

ANSWERS

A. 1

B. The same phase

C. The input resistance

D. Isolating loads

27 You can design an emitter follower circuit using the following steps:

1. Specify V_E. This is a DC voltage level, which is usually specified as half the supply voltage.

2. Find V_B. Use $V_B = V_E + 0.7$ volt.

3. Specify R_E. Often this is a given factor, especially if it is a motor or other load that is being driven.

4. Find I_E by using the following formula:
$$I_E = \frac{V_E}{R_E}$$

5. Find I_B by using the following formula:
$$I_B = \frac{I_E}{\beta}$$

6. Find I_2 by using $I_2 = 10I_B$.

7. Find R_2 by using the following formula:

$$R_2 = \frac{V_B}{I_2}$$

8. Find R_1 by using the following formula:

$$R_1 = \frac{V_S - V_B}{I_2 + I_B}$$

Usually, I_B is small enough to be dropped from this formula.

9. Choose the nearest standard values for R_1 and R_2.

10. Check that these standard values give a voltage close to V_B. Use the voltage divider formula.

A simple design example illustrates this procedure. Use the values shown in the circuit in Figure 8.26 for this problem.

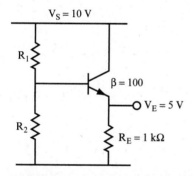

FIGURE 8.26

QUESTIONS

Work through Steps 1–10 to find the values of the two bias resistors.

1. $V_E =$ _____

2. $V_B =$ _____

3. $R_E =$ _____

4. $I_E =$ _____

5. $I_B =$ _____

6. $I_2 =$ _____

7. $R_2 =$ _____

8. $R_1 =$ _____

9. The nearest standard values are as follows:

$R_1 =$ _____

$R_2 =$ _____

10. $V_B =$ _____

ANSWERS

Your answers should be close to the following values:

1. 5 volts (This was given in Figure 8.26.)

2. 5.7 volts

3. 1 kΩ (This was given in Figure 8.26.)

4. 5 mA

5. 0.05 mA

6. 0.5 mA

7. 11.4 kΩ

8. 7.8 kΩ

9. The nearest standard values are 8.2 kΩ and 12 kΩ.

10. The standard resistor values result in $V_B = 5.94$ volts. This is a little higher than the V_B calculated in Step 2, but it is acceptable.

V_E is set by the biasing resistors. Therefore, it is not dependent upon the value of R_E. Almost any value of R_E can be used in this circuit. The minimum value for R_E is obtained by using this simple equation:

$$R_E = \frac{10\,R_2}{\beta}$$

ANALYZING AN AMPLIFIER

28 Up to now, the emphasis has been on designing a simple amplifier and an emitter follower. This section shows how to "analyze" a circuit that has already been designed.

In this case, to "analyze" means to calculate the collector DC voltage (the bias point) and find the AC gain. This procedure is basically the reverse of the design procedure.

Start with the circuit shown in Figure 8.27.

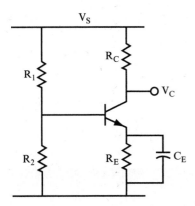

FIGURE 8.27

Following are the steps you use to analyze a circuit:

1. Find V_B by using the following equation:

$$V_B = V_S \times \frac{R_2}{R_1 + R_2}.$$

2. Find V_E by using $V_E = V_B - 0.7$ volt.

3. Find I_C by using the following equation:

$$I_C = \frac{V_E}{R_E}.$$

Note that $I_C = I_E$.

4. Find V_R by using $V_R = R_C \times I_C$.

5. Find V_C by using $V_C = V_S - V_R$. This is the bias point.

6. Find A_V by using the following equation:

$$A_V = \frac{R_C}{R_E}, \text{ or } A_V = \beta \times \frac{R_C}{R_{in}}$$

When you use the second formula, you must find the value of R_{in} (or h_{ie}) on the data sheets for the transistor from the manufacturer.

Use the circuit shown in Figure 8.28 for the following questions. For these questions, use $\beta = 100$, $R_{in} = 2 \text{ k}\Omega$ and the values given in the circuit drawing.

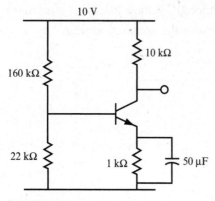

FIGURE 8.28

QUESTIONS

Calculate V_B, V_E, I_C, V_R, V_C, and A_V using Steps 1–6 of this problem.

1. $V_B =$ _____

2. $V_E =$ _____

3. $I_C =$ _____

4. $V_R =$ _____

5. $V_C =$ _____

6. $A_V =$ _____

ANSWERS

1. $V_B = 10 \times \dfrac{22\,k\Omega}{160\,k\Omega + 22\,k\Omega} = 1.2\,\text{volts}$

2. $V_E = 1.2 - 0.7 = 0.5\,\text{volt}$

3. $I_C = \dfrac{0.5\,V}{1\,k\Omega} = 0.5\,\text{mA}$

4. $V_R = 10\,k\Omega \times 0.5\,\text{mA} = 5\,\text{volts}$

5. $V_C = 10\,\text{volts} - 5\,\text{volts} = 5\,\text{volts}$ (This is the bias point.)

6. With the capacitor:

$$A_V = 100 \times \frac{10\,k\Omega}{2\,k\Omega} = 500\,(\text{a large gain})$$

Without the capacitor:

$$A_V = \frac{10\,k\Omega}{1\,k\Omega} = 10 \text{ (a small gain)}$$

29 You can determine the lowest frequency the amplifier will satisfactorily pass by following these simple steps:

1. Determine the value of R_E.
2. Calculate the frequency at which $X_C = R_E/10$. Use the capacitor reactance formula. (This is one of those "rules of thumb" that *can* be mathematically justified and gives reasonably accurate results in practice.)

QUESTIONS

For the circuit shown in Figure 8.28, find the following.

A. $R_E = $ _____

B. $f = $ _____

ANSWERS

A. $R_E = 1\,k\Omega$ (given in the circuit diagram)

B. So, you set $X_C = 100$ ohms, and use this formula:

$$X_C = \frac{1}{2\pi fC}$$

$$100\,\text{ohms} = \frac{0.16}{f \times 50\,\mu F} \quad \text{since } 0.16 = \frac{1}{2\pi}$$

So, the following is the result:

$$f = \frac{0.16}{100 \times 50 \times 10^{-6}} = 32\,Hz$$

30 For the circuit shown in Figure 8.29, follow the steps given in problems 28 and 29 to answer the following questions.

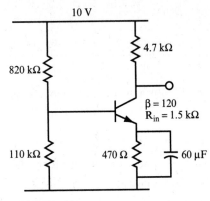

10 V

4.7 kΩ

820 kΩ

β = 120
R_{in} = 1.5 kΩ

110 kΩ   470 Ω   60 μF

FIGURE 8.29

QUESTIONS

1. V_B = _____

2. V_E = _____

3. I_C = _____

4. V_R = _____

5. V_C = _____

6. With capacitor:

A_V = _____

Without capacitor:

A_V = _____

7. Low frequency check:

f = _____

ANSWERS

Your answers should be close to these.

1. 1.18 volts

2. 0.48 volts

3. 1 mA

4. 4.7 volts

5. 5.3 volts (bias point)

6. With capacitor: 376

Without capacitor: 10

7. 57 Hz (approximately)

THE JFET AS AN AMPLIFIER

31 Chapter 3 discussed the JFET in problems 28–31, and Chapter 4 discussed the JFET in problems 37–41. You may want to review these problems before answering the questions in this problem. Figure 8.30 shows a typical biasing circuit for a JFET.

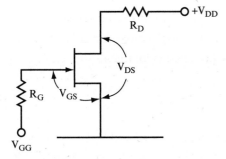

FIGURE 8.30

QUESTIONS

A. What type of JFET is depicted in the circuit?_____

_____

B. What value of V_{GS} would you need to turn the JFET completely ON?_____

_____

C. What drain current flows when the JFET is completely ON?_____

_____

D. What value of V_{GS} would you need to turn the JFET completely OFF?_____

_____

E. When a JFET is alternately turned completely ON and OFF in a circuit, what type of component are you using the JFET as?_____

_____

ANSWERS

A. N-channel JFET.

B. $V_{GS} = 0$ V to turn the JFET completely ON.

C. Drain saturation current (I_{DSS}).

D. V_{GS} should be a negative voltage for the N-channel JFET to turn it completely OFF. The voltage must be larger than or equal to the cutoff voltage.

E. The JFET is being used as a switch.

32 You can use a JFET to amplify AC signals by biasing the JFET with a gate to source voltage about halfway between the ON and OFF states. You can find the drain current that flows in a JFET biased to a particular V_{GS} by using the following equation for the transfer curve:

$$I_D = I_{DSS}\left(1 - \frac{V_{GS}}{V_{GS(off)}}\right)^2$$

In this equation, I_{DSS} is the value of the drain saturation current, and $V_{GS(off)}$ is the gate to source voltage at cutoff. Both of these are indicated on the transfer curve shown in Figure 8.31.

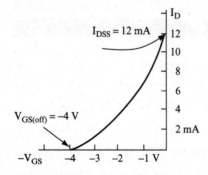

FIGURE 8.31

For the transfer curve shown in Figure 8.31, I_{DSS} = 12 mA and $V_{GS(off)}$ = −4 volts. Setting the bias voltage at V_{GS} = −2 volts returns the following value for the drain current:

$$I_D = 12\,mA \times \left(1 - \frac{-2}{-4}\right)^2 = 12\,mA = (0.5)^2 = 3\,mA$$

QUESTIONS

Calculate the drain current for the following:

A. V_{GS} = −1.5 _____

_____

B. V_{GS} = −0.5 volts_____

_____

ANSWERS

A. 4.7 mA

B. 9.2 mA

NOTE   Data sheets give a wide range of possible I_{DSS} and $V_{GS(off)}$ values for a given JFET. You may need to resort to actually measuring these with the method shown in Project 4-2.

33 For the circuit shown in Figure 8.30, you choose the value of the drain to source voltage, V_{DS}, and then calculate the value of the load resistor, R_D, by using the following equation:

$$R_D = \frac{(V_{DD} - V_{DS})}{I_D}$$

For this problem, use I_D = 3 mA, and a drain supply voltage (V_{DD}) of 24 volts. Calculate the value of R_D that results in the specified value of V_{DS}; this is also the DC output voltage of the amplifier.

QUESTION

Calculate the value of R_D that will result in V_{DS} = 10 volts. _____

_____

ANSWER

$$R_D = \frac{(V_{DD} - V_{DS})}{I_D} = \frac{(24\,\text{volts} = 10\,\text{volts})}{3\,\text{mA}} = \frac{14\,\text{volts}}{3\,\text{mA}} = 4.67\,\text{k}\Omega$$

---

34 The circuit shown in Figure 8.32 (which is referred to as a *JFET common source amplifier*) applies a 0.5 V$_{pp}$ sine wave to the gate of the JFET and produces an amplified sine wave output from the drain.

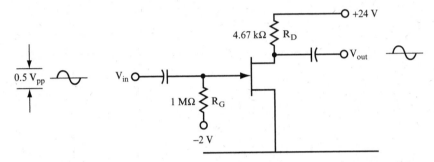

FIGURE 8.32

The input sine wave is added to the −2 volt bias applied to the gate of the JFET. Therefore, V_{GS} varies from −1.75 to −2.25 volts.

QUESTION

Using the formula in problem 32, calculate I_D for the maximum and minimum values of V_{GS}.

ANSWER

For $V_{GS} = -1.75$ volts, $I_D = 3.8$ mA
For $V_{GS} = -2.25$ volts, $I_D = 2.3$ mA

---

35 As the drain current changes, V_{RD} (the voltage drop across resistor R_D) also changes.

QUESTION

For the circuit shown in Figure 8.32, calculate the values of V_{RD} for the maximum and minimum values of I_D you calculated in problem 34. _____

ANSWER

For $I_D = 3.8\,\text{mA}$, $V_{RD} = 3.8\,\text{mA} \times 4.67\,\text{k}\Omega = 17.7\,\text{volts}$

For $I_D = 2.3\,\text{mA}$, $V_{RD} = 2.3\,\text{mA} = 4.67\,\text{k}\Omega = 10.7\,\text{volts}$

This corresponds to a 7 V_{pp} sine wave.

36 As the voltage drop across R_D changes, the output voltage also changes.

QUESTION

For the circuit shown in Figure 8.32, calculate the values of V_{out} for the maximum and minimum values of V_{RD} you calculated in problem 35. _____

_____

ANSWER

For $V_{RD} = 17.7\,\text{volts}$, $V_{out} = V_{DD} - V_{RD} = 24\,\text{volts} - 17.7\,\text{volts} = 6.3\,\text{volts}$

For $V_{RD} = 10.7\,\text{volts}$, $V_{out} = V_{DD} - V_{RD} = 24\,\text{volts} = 10.7\,\text{volts} = 13.3\,\text{volts}$

Therefore, the output signal is a 7 V_{pp} sine wave.

37 Table 8-1 shows the results of the calculations made in problems 34–36 including the DC bias point.

TABLE 8-1: Calculation Results

V_{GS}	I_D	V_{RD}	V_{out}
−1.75 volts	3.8 mA	17.7 volts	6.3 volts
−2.0 volts	3.0 mA	14.0 volts	10.0 volts
−2.25 volts	2.3 mA	10.7 volts	13.3 volts

QUESTION

What are some characteristics of the AC output signal?_____

_____

ANSWER

The output signal is a 7 V_{pp} sine wave with the same frequency as the input sine wave. As the input voltage on V_{GS} increases (toward 0 volts), the output decreases. As the input voltage decreases (becomes more negative), the output voltage increases. This means that the output is 180 degrees out of phase with the input.

38 You can calculate the AC voltage gain for the amplifier discussed in problems 34–37 by using the following formula:

$$A_v = \frac{-V_{out}}{V_{in}}$$

The negative sign in this formula indicates that the output signal is 180 degrees out of phase from the input signal.

QUESTION

Calculate the AC voltage gain for the amplifier discussed in problems 34–37. _____

ANSWER

$$A_V = \frac{-7\,V_{pp}}{0.5\,V_{pp}} = -14$$

39 You can also calculate the AC voltage gain by using the following formula:

$$A_v = -(g_m)(R_D)$$

In this equation, g_m is the *transconductance* and is a property of the JFET. It is also called the *forward transfer admittance*. A typical value for g_m is usually provided for JFETs in the data sheet from the manufacturers. You can also use the data in Table 8-1 to calculate g_m using the following formula:

$$g_m = \frac{\Delta I_D}{\Delta V_{GS}}$$

In this equation, Δ indicates the change or variation in V_{GS} and the corresponding drain current. The unit for transconductance is mhos.

QUESTIONS

A. Using the data from Table 8-1, what is the value of g_m for the JFET used in the amplifier? _____

_____

B. What is the corresponding AC voltage gain? _____

_____

ANSWERS

A. $gm = \dfrac{1.5\,mA}{0.5\,V} = 0.003$ mhos

B. $A_v = -(0.003)(4670) = -14$, the same result you found in problem 38

40 Design a JFET common source amplifier using a JFET with $I_{DSS} = 14.8$ mA and $V_{GS(off)} = -3.2$ volts. The input signal is 40 mV_{pp}. The drain supply is 24 volts.

QUESTIONS

A. Determine the value of V_{GS} that will bias the JFET at a voltage near the middle of the transfer curve. _____

_____

B. Calculate the drain current when V_{GS} is at the value determined in step A, using the formula in problem 32. _____

_____

C. Choose a value of V_{DS} and calculate the value of R_D using the formula in problem 33. _____

_____

D. Calculate the maximum and minimum values of V_{GS} that result from the input signal, and the corresponding values of drain current using the procedure in problem 34. _____

_____

E. Calculate the maximum and minimum values of V_{out} that result from the input signal using the procedures in problems 35 and 36. _____

_____

F. Calculate the gain of the amplifier. _____

_____

ANSWERS

A. $V_{GS} = -1.6$ volts

B. $I_D = 3.7$ mA

C. For $V_{DS} = 10$ volts,

$$R_D = \frac{14\,\text{volts}}{3.7\,\text{mA}} = 3780 \text{ ohms}$$

D. V_{GS} will vary from -1.58 to -1.62 volts. Use the formula to calculate values of drain current. I_D will vary from 3.79 to 3.61 mA.

E. V_{RD} will vary from 14.3 to 13.6 volts. Therefore, V_{out} will vary from 9.7 to 10.4 volts.

F. $A_v = \dfrac{-0.7}{0.04} = -17.5$

41 Use the results of problem 40, question D, to answer the following question.

QUESTIONS

Calculate the transconductance of the JFET and the AC voltage gain using the formulas discussed in problem 39. _____

_____

ANSWERS

$$g_m = \frac{\Delta I_D}{\Delta V_{GS}} = \frac{0.18\,\text{mA}}{40\,\text{mV}} = 0.0045 \text{ mhos}$$

$$A_v = -(g_m)(R_D) = -(0.0045)(3780) = -17$$

This is close to the value you found in problem 40, question F.

42 Figure 8.33 shows a JFET amplifier circuit that uses one power supply, rather than separate power supplies for the drain and gate used in the amplifier discussed in problems 34–41.

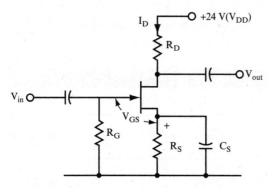

FIGURE 8.33

The DC voltage level of the gate is zero because the gate is tied to ground through R_G. Therefore, the voltage drop across R_S becomes the gate to source voltage. To design the circuit, you must find values for both R_S and R_D. Use the same bias point for this problem as you used for the amplifier discussed in problems 34–41: $V_{GS} = -2$ volts and $I_D = 3$ mA. Follow these steps:

1. Calculate R_S, using the following formula, recognizing that $V_{RS} = |V_{GS}|$:

$$R_S = \frac{V_{RS}}{I_D} = \frac{V_{GS}}{I_D}$$

2. Calculate R_D using the following formula, using $V_{DS} = 10$ volts, the same value you used for the amplifier discussed in problems 34–41:

$$R_D = \frac{(V_{DD} - V_{DS} - V_S)}{I_D}$$

3. Calculate X_{CS} using the following formula:

$$X_{CS} = \frac{R_S}{10}$$

Then, calculate C_S using the following formula:

$$X_{CS} = \frac{1}{2\pi f C_S}$$

4. Calculate the peak-to-peak output voltage using the procedures shown in problems 34–36.

5. Calculate the AC voltage gain using this formula:

$$A_V = \frac{-V_{out}}{V_{in}}$$

NOTE Choose the value of C_S so that its reactance is less than 10 percent of R_S at the lowest frequency you need to amplify. The DC load for the JFET is R_D plus R_S. The AC load is R_D only because C_S bypasses the AC signal around R_S, which keeps the DC operating point stable. The use of C_S reduces the gain slightly because you now use a smaller R_D to calculate the AC voltage swings at the output.

QUESTIONS

A. What is the value of R_S? _____

_____

B. What is the value of R_D? _____

_____

C. What is the value of C_S? Assume $f = 1$ kHz. _____

_____

D. Calculate the peak-to-peak V_{out} for $V_{in} = 0.5$ V_{pp}. _____

_____

E. What is the voltage gain? _____

_____

ANSWERS

A. $R_S = \dfrac{2\,\text{volts}}{3\,\text{mA}} = 667\,\text{ohms}$

B. $R_D = \dfrac{12\,\text{volts}}{3\,\text{mA}} = 4\,\text{k}\Omega$

C. $X_{CS} = 66.7$ ohms, $C_S = 2.4\ \mu\text{F}$

D. The AC drain current will still vary from 3.8 to 2.3 mA, as in problem 37. The voltage across R_D is now 6 V_{pp} because R_D is 4 kΩ. The output voltage is also 6 V_{pp}.

E. $A_V = \dfrac{-6}{0.5} = -12$ The gain is 12.

THE OPERATIONAL AMPLIFIER

43 The operational amplifier (op-amp) in use today is actually an integrated circuit (IC). This means that the device has numerous transistors and other components constructed on a small silicon chip. These IC op-amps are much smaller and, therefore, more practical than an amplifier with equivalent performance that is made with discrete components.

You can purchase op-amps in different case configurations. Some of these configurations are the Transistor Outline (TO) metal package, the flat pack, and the dual in-line pin (DIP) package. You can also find two op-amps (dual) or four op-amps (quad) in a single IC.

Their size, low cost, and wide range of applications have made op-amps so common today that they are thought of as a circuit device or component in and of themselves, even though a typical op-amp may contain 20 or more transistors in its design. The characteristics of op-amps closely resemble those of an ideal amplifier. Following are these characteristics:

- High input impedance (does not require input current)

- High gain (used for amplifying small signal levels)

- Low output impedance (not affected by the load)

QUESTIONS

A. What are the advantages of using op-amps? _____

_____

B. Why are op-amps manufactured using IC techniques? _____

_____

ANSWERS

A. Small size, low cost, wide range of applications, high input impedance, high gain, and low output impedance.

B. Because of the large numbers of transistors and components that are required in the design of an op-amp, they must be constructed on a single, small silicon chip using IC manufacturing techniques to be of a reasonable size.

44 Figure 8.34 shows the schematic symbol for an op-amp.

An input at the inverting input results in an output that is 180 degrees out of phase with the input. An input at the noninverting input results in an output that is in phase with the input. Both positive and negative voltage supplies are required, and the data sheet will specify their values for the particular op-amp you use. Datasheets usually contain circuit diagrams showing how you should connect external components to the op-amp for specific applications. These circuit diagrams (showing how a particular op-amp can be used for various applications) can be useful to the designer or the hobbyist.

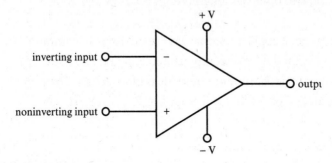

FIGURE 8.34

QUESTIONS

A. How many terminals does the op-amp require, and what are their functions? _____
_____

B. How is the output related to the input when the input is connected to the inverting input? _____
_____

ANSWERS

A. Five—two input terminals, one output terminal, two power supply terminals.

B. The output is 180 degrees out of phase with the input.

45 Figure 8.35 shows a basic op-amp circuit. The input signal is connected to an inverting input, as indicated by the negative sign. Therefore, the output signal will be 180 degrees out of phase with the input.

You can find the AC voltage gain for the circuit using the following equation:

$$A_v = \frac{-R_F}{R_{in}}$$

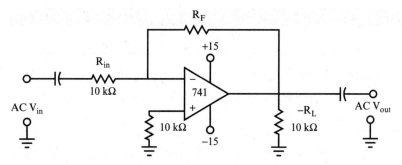

FIGURE 8.35

Resistor R_F is called a *feedback resistor* because it forms a feedback path from the output to the input. Many op-amp circuits use a feedback loop. Because the op-amp has such a high gain, it is easy to saturate it (at maximum gain) with small voltage differences between the two input terminals. The feedback loop allows the operation of the op-amp at lower gains, allowing a wider range of input voltages. When designing a circuit, you can choose the value of the feedback resistor to achieve a specific voltage gain. The role of the capacitors in the diagram is to block DC voltages.

QUESTIONS

A. Calculate the value of R_F that would give the amplifier an AC voltage gain of 120. _

_____

B. Calculate $A_C V_{out}$ if AC V_{in} is 5 mV$_{rms}$. _____

_____

ANSWERS

A. $R_F = 120 \times 10\,k\Omega = 1.2\,M\Omega$

B. $V_{out} = 120 \times 5\,mV = 0.6\,V_{rms}$

The output signal is inverted with respect to the input signal.

46 Use the op-amp circuit shown in Figure 8.35 to build an amplifier with an output voltage of 12 V$_{pp}$, an AC voltage gain of 50, and with $R_{in} = 6.8\,k\Omega$.

QUESTIONS

A. Calculate the value of R_F. _____

_____

B. Calculate the value of V_{in} required to produce the output voltage specified earlier. _

_____

ANSWERS

A. $R_F = 50 \times 6.8\,k\Omega = 340\,k\Omega$

B. $V_{in} = \dfrac{12\,V_{pp}}{50} = 0.24\,V_{pp}$ or $0.168\,V_{rms}$

PROJECT 8.2: The Operational Amplifier

OBJECTIVE

The objective of this project is to demonstrate how AC voltage gain changes when you use feedback resistors of different values in an op-amp circuit.

GENERAL INSTRUCTIONS

After the circuit is set up, you measure V_{out} for each value of R_F, and find A_V, using the ratio V_{out}/V_{in}. You also determine a calculated A_V using the ratio R_F/R_{in} in each case to determine how close the calculated A_V is to the measured A_V.

Parts List

You need the following equipment and supplies:

❑ One 0.1 µF capacitor.

❑ Two 10 kΩ, 0.25-watt resistors.

❑ One 51 kΩ, 0.25-watt resistor.

❑ One 100 kΩ, 0.25-watt resistor.

❑ One 150 kΩ, 0.25-watt resistor.

- One 220 kΩ, 0.25-watt resistor.
- One 270 kΩ, 0.25-watt resistor.
- One 330 kΩ, 0.25-watt resistor.
- One 380 kΩ, 0.25-watt resistor.
- Two terminal blocks.
- Two 6-volt battery packs (4 AA batteries each).
- One function generator.
- One oscilloscope.
- One breadboard.
- One OPA134 operational amplifier. This op-amp comes in a few different packages; get the 8-pin dual in-line (DIP) version. Figure 8.36 shows the pinout diagram for the OPA134. When you insert the op-amp into the breadboard, try not to bend any of the leads. The leads on dual in-line packages are fragile and will break off if you bend them more than once or twice.

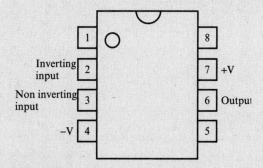

FIGURE 8.36

STEP-BY-STEP INSTRUCTIONS

Set up the circuit shown in Figure 8.37 using the 51 kΩ resistor for R_F. Figure 8.38 shows the battery connections. If you have some experience in building circuits, this schematic (along with the previous parts list) should provide all the information you need to build the circuit. If you need a bit more help, look at the photos of the completed circuit in the "Expected Results" section. One unusual aspect of this circuit you may want to look for in the photos is how the −V bus of one 6-volt battery pack should be connected to the +V bus of the other 6-volt battery pack.

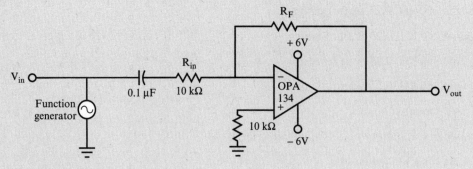

FIGURE 8.37

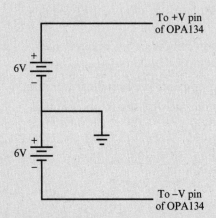

FIGURE 8.38

Carefully check your circuit against the diagram.

After you check your circuit, follow these steps, and record your measurements in the blank table following the steps.

1. Connect the oscilloscope probe for channel 2 to a jumper wire connected to V_{in}. Connect the ground clip to a jumper wire attached to the ground bus.

2. Connect the oscilloscope probe for channel 1 to a jumper wire connected to V_{out}, and then connect the ground clip to a jumper wire attached to the ground bus.

3. Set the function generator to generate a 10 kHz sine wave with approximately 0.2 V_{pp}.

4. Measure and record V_{out} and V_{in}.

5. Change the feedback resistor to the value shown in the next row of the table (labeled 100 kΩ in this instance). Each time you change the resistor, it's advisable to disconnect the batteries to avoid shorting wires.

6. Measure and record V_{out} and V_{in}.

7. Repeat steps 5 and 6 until you have recorded V_{out} and V_{in} for the last row of the table.

R_F	Calculated A_V (R_F/R_{in})	V_{in} (volts)	V_{out} (volts)	Measured A_V (V_{out}/V_{in})
51 kΩ				
100 kΩ				
150 kΩ				
220 kΩ				
270 kΩ				
330 kΩ				
380 kΩ				

8. Determine the calculated A_V and the measured A_V, and record these values in each row of the table.

EXPECTED RESULTS

Figure 8.39 shows the breadboarded circuit for this project.

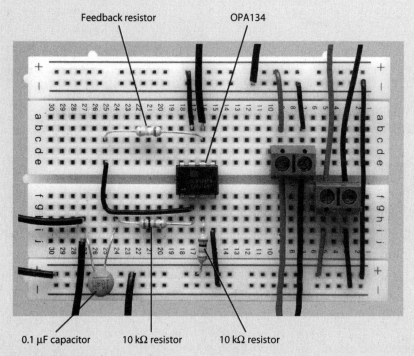

Feedback resistor   OPA134

0.1 μF capacitor   10 kΩ resistor   10 kΩ resistor

FIGURE 8.39

Figure 8.40 shows a function generator and oscilloscope attached to the circuit.

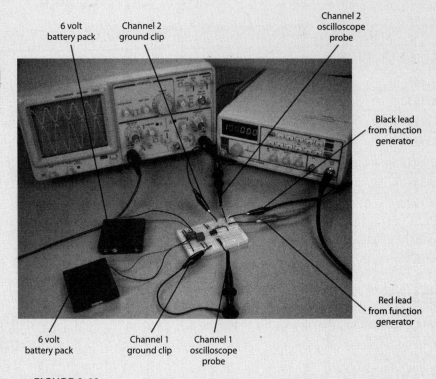

FIGURE 8.40

The input signal is represented by the upper sine wave shown in Figure 8.41, and the output signal is represented by the lower sine wave. Count the number of divisions for the peak-to-peak output sine wave, and multiply that number by the corresponding VOLTS/DIV setting to determine V_{out} and V_{in}.

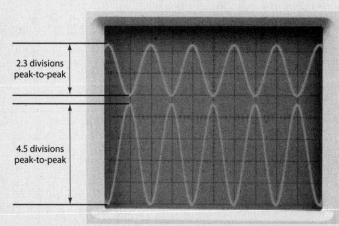

FIGURE 8.41

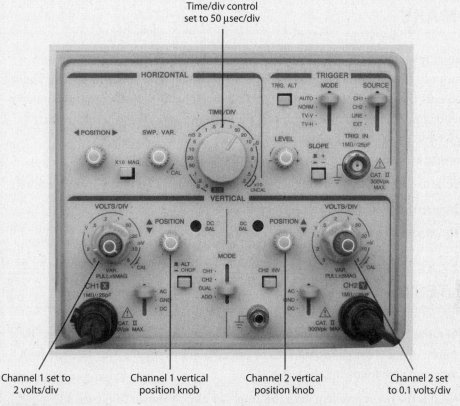

Time/div control
set to 50 μsec/div

Channel 1 set to
2 volts/div

Channel 1 vertical
position knob

Channel 2 vertical
position knob

Channel 2 set
to 0.1 volts/div

FIGURE 8.42

As you measure V_{in} and V_{out}, you may need to adjust the TIME/DIV control, the VOLTS/DIV control, and vertical POSITION controls on the oscilloscope. The controls shown in Figure 8.42 are adjusted to measure V_{out} when $R_F = 380$ kΩ.

Your values should be close to those shown in the following table.

R_F	Calculated A_V (R_F/R_{in})	V_{in} (volts)	V_{out} (volts)	Measured A_V (V_{out}/V_{in})
51 kΩ	5.1	0.23	1.2 volts	5.2
100 kΩ	10	0.23	2.3 volts	10
150 kΩ	15	0.23	3.5 volts	15.2
220 kΩ	22	0.23	5.1 volts	22.2
270 kΩ	27	0.23	6.2 volts	27
330 kΩ	33	0.23	7.6 volts	33
380 kΩ	38	0.23	9 volts	39

The measured values of A_V are quite close to the calculated values of A_V, well within variations that could be caused by the ± 5 percent tolerance specified for resistor values.

SUMMARY

This chapter introduced the most common types of amplifiers in use today: the common emitter BJT, the common source JFET, and the op-amp. At best, this chapter has scratched only the surface of the world of amplifiers. Actually, there are many variations and types of amplifiers. Still, the terminology and design approach you learned here should give you a basic foundation for further study.

Following are the key skills you gained in this chapter:

- How to design a simple amplifier when the bias point and the gain are specified

- How to do the same for an emitter follower

- How to analyze a simple amplifier circuit

SELF-TEST

These questions test your understanding of the material presented in this chapter. Use a separate sheet of paper for your diagrams or calculations. Compare your answers with the answers provided following the test.

1. What is the main problem with the amplifier circuit shown in Figure 8.1? _____

    _____

2. What is the gain formula for that circuit? _____

    _____

3. Does it have a high or low gain? _____

    _____

Use the circuit shown in Figure 8.43 for questions 4–8.

4. Design an amplifier so that the bias point is 5 volts, and the AC voltage gain is 15. Assume $\beta = 75$, $R_{in} = 1.5$ kΩ, $V_S = 10$ volts, and $R_C = 2.4$ kΩ. Add capacitor C_E to the circuit and calculate a suitable value to maintain maximum AC voltage gain at 50 Hz. What is the approximate value of this gain?_____

    _____

5. Repeat question 4 with these values: $V_S = 28$ volts, $\beta = 80$, $R_{in} = 1$ kΩ, and $R_C = 10$ kΩ. The bias point should be 14 volts and the AC voltage gain 20._____

    _____

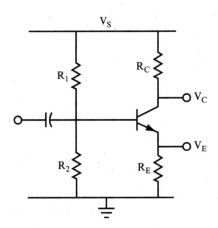

FIGURE 8.43

6. Repeat question 4 with these values: V_S = 14 volts, β = 250, R_{in} = 1 kΩ, and R_C = 15 kΩ. The bias point should be 7 volts and the AC voltage gain 50._____

_____

7. Design an emitter follower amplifier given that V_S = 12 volts, R_E = 100 ohms, β = 35, V_E = 7 volts, and R_C = 0 ohms._____

_____

8. Design an emitter follower amplifier given that V_S = 28 volts, R_E = 100 ohms, β = 35, V_E = 7 volts, and R_C = 0 ohms. _____

_____

In questions 9–11, the resistance and β values are given. Analyze the circuit to find the bias point and the gain.

9. R_1 = 16 kΩ, R_2 = 2.2 kΩ, R_E = 100 ohms, R_C = 1 kΩ, β = 100, and V_S = 10 volts_

_____

10. R_1 = 36 kΩ, R_2 = 3.3 kΩ, R_E = 110 ohms, R_C = 2.2 kΩ, β = 50, and V_S = 12 volts_

_____

11. R_1 = 2.2 kΩ, R_2 = 90 kΩ, R_E = 20 ohms, R_C = 300 kΩ, β = 30, and V_S = 50 volts_

_____

12. The circuits from questions 4 and 5 are connected to form a two-stage amplifier. What is the gain when there is an emitter bypass capacitor for both transistors? When the capacitor is not used in either of them?_____

_____

13. Design a JFET amplifier using the circuit shown in Figure 8.32. The characteristics of the JFET are I_{DSS} = 20 mA and $V_{GS(off)}$ = −4.2 volts. The desired value of V_{DS} is 14 volts. Find the value of R_D._____

_____

14. If the transconductance of the JFET used in question 13 is 0.0048 mhos, what is the voltage gain?_____

_____

15. If the desired output is 8 V_{pp} for the JFET of questions 13 and 14, what should the input be?_____

_____

16. Design a JFET amplifier using the circuit in Figure 8.33. The JFET characteristics are I_{DSS} = 16 mA and $V_{GS(off)}$ = −2.8 volts. Using a V_{DS} of 10 volts, find the values of R_S, C_S, and R_D._____

_____

17. If the input to the JFET in question 16 is 20 mV_{pp}, what is the AC output voltage, and what is the gain?_____

_____

18. For the op-amp circuit shown in Figure 8.35, what is the output voltage if the input is 50 mV and the feedback resistor is 750 kΩ?_____

_____

ANSWERS TO SELF-TEST

If your answers do not agree with those provided here, review the problems in parentheses before you go on to Chapter 9, "Oscillators."

1.	Its bias point is unstable, and its gain varies with temperature. Also, you cannot guarantee what the gain will be.	(problem 10)
2.	$A_v = \beta \times \dfrac{R_c}{R_{in}}$	(problem 10)
3.	Usually the gain is quite high.	(problem 10)
	For Numbers 4–6, suitable values are given. Yours should be close to these.	

4.	$R_1 = 29\ k\Omega$, $R_2 = 3.82\ k\Omega$, $R_E = 160$ ohms, $C_E = 200\ \mu F$, $A_V = 120$	(problems 17–22)
5.	$R_1 = 138\ k\Omega$, $R_2 = 8\ k\Omega$, $R_E = 500$ ohms, $C_E = 64\ \mu F$, $A_V = 800$	(problems 17–22)
6.	$R_1 = 640\ k\Omega$, $R_2 = 45\ k\Omega$, $R_E = 300$ ohms, $C_E = 107\ \mu F$, $A_V = 750$	(problems 17–22)
7.	$R_1 = 8\ k\Omega$; $R_2 = 11.2\ k\Omega$	(problem 27)
8.	$R_1 = 922$ ohms; $R_2 = 385$ ohms	(problem 27)
9.	$V_C = 5$ volts, $A_V = 10$	(problems 28–30)
10.	$V_C = 6$ volts, $A_V = 20$	(problems 28–30)
11.	$V_C = 30$ volts, $A_V = 15$	(problems 28–30)
12.	When the capacitor is used, $A_V = 120 \times 800 = 96{,}000$.	(problems 17–22)
	When the capacitor is not used, $A_V = 15 \times 20 = 300$.	
13.	Use $V_{GS} = -2.1$ volts, then $I_D = 5$ mA, $R_D = 2\ k\Omega$.	(problems 31–33)
14.	$A_v = -9.6\ mV_{pp}$	(problem 39)
15.	$V_{in} = 83\ mV_{pp}$	(problem 38)
16.	Use $V_{GS} = -1.4$ volts, then $I_D = 4$ mA.	(problem 42)
	$R_S = 350$ ohms	
	$C_S = 4.5\ \mu F$ (assume $f = 1$ kHz)	
	$R_D = 3.15\ k\Omega$	
17.	V_{GS} varies from -1.39 to -1.41 volts, I_D varies from 4.06 to 3.94 mA, V_{out} will be 400 mV_{pp}, $$A_v = \frac{-400}{20} = -20$$	(problem 42)
18.	$A_v = -75$, $V_{out} = 3.75$ volts	(problem 45)

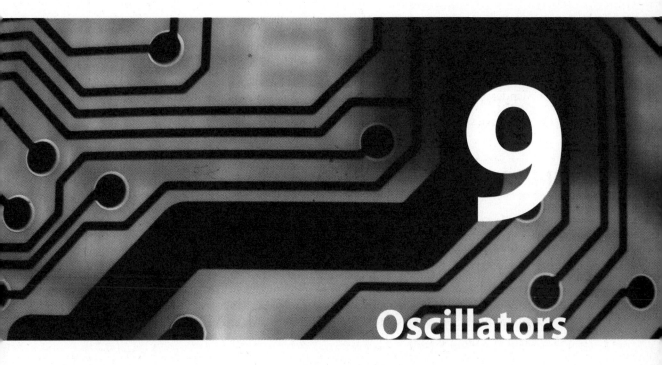

9

Oscillators

This chapter introduces you to oscillators. An *oscillator* is a circuit that produces a continuous output signal. There are many types of oscillator circuits used extensively in electronic devices. Oscillators can produce a variety of different output signals, such as sine waves, square waves, or triangle waves.

When the output signal of an oscillator is a sine wave of constant frequency, the circuit is called a *sine wave oscillator*. Radio and television signals are sine waves transmitted through the air, and the 120-volts AC from the wall plug is a sine wave, as are many test signals used in electronics.

This chapter introduces three basic sine wave oscillators. They all rely on resonant LC circuits as described in Chapter 7, "Resonant Circuits," to set the frequency of the sine wave.

When you complete this chapter, you will be able to do the following:

- Recognize the main elements of an oscillator.

- Differentiate between positive and negative feedback.
- Specify the type of feedback that causes a circuit to oscillate.
- Specify at least two methods of obtaining feedback in an oscillator circuit.
- Understand how resonant LC circuits set the frequency of an oscillator.
- Design a simple oscillator circuit.

UNDERSTANDING OSCILLATORS

1 An oscillator can be divided into three definite sections:

- An amplifier
- The feedback connections
- The components that set frequency

The amplifier replaces the switch in the basic oscillator circuit, introduced in problem 35 of Chapter 7 (Figure 7.46).

QUESTION

Draw an oscillator circuit, and label the parts. Use a separate sheet of paper for your drawing.

ANSWER

See Figure 9.1.

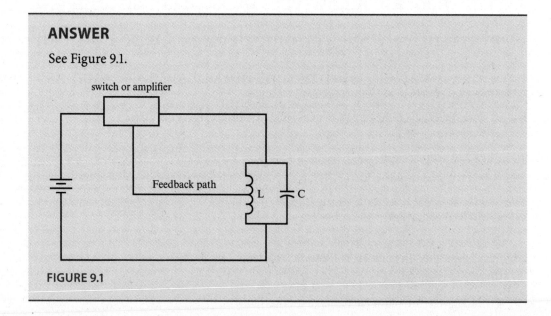

FIGURE 9.1

2 When you connect the output of an amplifier to its input, you get feedback. If the feedback is "out of phase" with the input, as shown in Figure 9.2, then the feedback is *negative*.

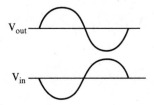

FIGURE 9.2

When the signal from the collector is fed back to the base of the transistor through a feedback resistor (R_f), as in the circuit shown in Figure 9.3, the feedback signal is out of phase with the input signal. Therefore, the feedback is negative.

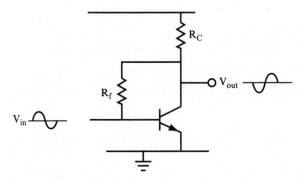

FIGURE 9.3

Negative feedback is used to stabilize the operation of an amplifier by doing the following:

- Preventing the DC bias point and gain of an amplifier from being affected by changes in temperature
- Reducing distortion in amplifiers, thereby improving the quality of the sound

QUESTIONS

A. Why would feedback signals be used in quality audio amplifiers? _____

_____

B. What kind of feedback do they have?_____

_____

3 If the feedback from the output is in phase with the input, as shown in Figure 9.4, the circuit's feedback is *positive*.

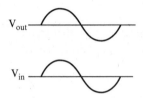

FIGURE 9.4

In the circuit shown in Figure 9.5, the collector of the second transistor is connected to the base of the first transistor. Because the output signal at the collector of the second transistor is in phase with the input signal at the base of the first transistor, this circuit has positive feedback.

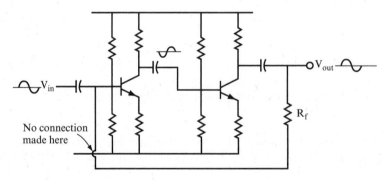

FIGURE 9.5

Positive feedback can cause an amplifier to oscillate even when there is *no external input*.

QUESTIONS

A. What type of feedback is used to stabilize an amplifier? _____
_____

B. What type of feedback is used in oscillators? _____

_____

C. What parts of an amplifier do you connect to produce feedback? _____

_____

ANSWERS

A. Negative feedback.

B. Positive feedback.

C. Connect the output of an amplifier to its input.

4 The amplifier shown in Figure 9.6 is the same type of amplifier that was discussed in problems 11–18 of Chapter 8, "Transistor Amplifiers." It is called a *common emitter amplifier*.

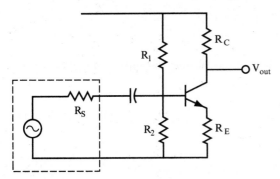

FIGURE 9.6

QUESTIONS

A. What effect would negative feedback have on this amplifier? _____

_____

B. What effect would positive feedback have on this amplifier? _____

_____

ANSWERS

A. Stabilize it, reduce gain, and reduce distortion

B. Cause it to oscillate

5 In the circuit shown in Figure 9.6, an input signal applied to the base will be amplified.

QUESTIONS

A. What is the basic formula for an amplifier's voltage gain? _____

_____

B. What is the voltage gain formula for the amplifier circuit shown in Figure 9.6? ____

_____

ANSWERS

A. $A_V = \beta \times \dfrac{R_L}{R_{in}}$

B. $A_V = \dfrac{R_L}{R_E} = \dfrac{R_C}{R_E}$

(as discussed in problem 12 of Chapter 8)

6 In the circuit shown in Figure 9.7, an input signal is applied to the emitter of the transistor instead of the base. This circuit is called a *common base amplifier*.

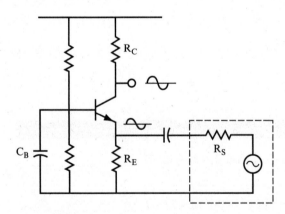

FIGURE 9.7

> **NOTE** When you apply a signal to the emitter, it changes the voltage drop across the base-emitter diode, just as an input signal applied to the base does. Therefore, a signal applied to the emitter changes the base current and the collector current, just as if you had applied a signal to the base.

The voltage gain formula for this type of amplifier can be simplified because the input impedance to the amplifier is so low when the signal is fed into the emitter that you can discount it. This results in the following voltage gain formula for the common base amplifier:

$$A_V = \frac{R_L}{R_S}$$

R_S is the output resistance or impedance of the source or generator. It is also called the *internal impedance* of the source.

QUESTION

What is the voltage gain formula for the circuit shown in Figure 9.7? _____

_____

ANSWER

$$A_V = \frac{R_L}{R_S} = \frac{R_C}{R_S}, (R_C \text{ is the load in this circuit.})$$

7 Notice that the input and output sine waves in Figure 9.7 are in phase. Although the signal is amplified, it is not inverted.

QUESTIONS

A. What happens to the input signal to the amplifier when you apply it to the emitter instead of the base? _____

_____

B. Is the input impedance of the common base amplifier high or low compared to the common emitter amplifier? _____

_____

C. What is the gain formula for the common base amplifier? _____

_____

ANSWERS

A. Amplified and not inverted

B. Low

C. $A_V = \dfrac{R_L}{R_S} = \dfrac{R_C}{R_S}$

8 Figure 9.8 shows an amplifier circuit with a parallel inductor and capacitor connected between the collector of the transistor and ground. A parallel inductor and capacitor circuit is sometimes called a *tuned* (or *resonant*) *load*.

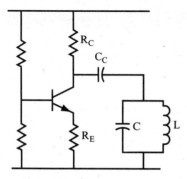

FIGURE 9.8

In this circuit, the inductor has a small DC resistance, which could pull the collector DC voltage down to near 0 volts. Therefore, you include capacitor C_c in the circuit to allow AC signals to pass through the LC circuit while preventing the collector DC voltage from being pulled down to 0 volts.

QUESTIONS

A. What term would you use to describe the load in this circuit? _____

_____

B. Does the circuit contain all three components of an oscillator at this point? _____

_____

ANSWERS

A. Resonant or tuned.

B. No, the feedback connections are missing.

NOTE The circuit shown in Figure 9.8 does not have an input signal either to the emitter or to the base. By adding a feedback connection to a parallel LC circuit, you provide an input signal to the emitter or base, as explained later in this chapter.

9 Write the voltage gain formulas for the following circuits. Refer to the circuits and voltage gain formulas in problems 4–6, if necessary.

QUESTIONS

A. Common emitter circuit _____

_____

B. Common base circuit _____

_____

ANSWERS

A. $A_V = \dfrac{R_C}{R_E}$

B. $A_V = \dfrac{R_C}{R_S}$

10 You can use common emitter and common base amplifier circuits in oscillators, and in each case, you would usually also include an extra capacitor.

In a common emitter amplifier, you can add a capacitor (C_E) between the emitter and ground, as discussed in problems 19 and 20 of Chapter 8.

In a common base circuit, you can add a capacitor (C_B) between the base and the ground, as is shown in Figure 9.7.

QUESTION

What is the general effect in both cases? _____

_____

ANSWER

An increase in the gain of the amplifier

The gain is increased to the point where you can consider it "large enough" to use the amplifier as an oscillator. When these capacitors are used in either a common emitter or common base amplifier, it is not usually necessary to calculate the gain of the amplifier.

11 An LC circuit has a resonance frequency that you can determine using the methods discussed in problems 6–12 of Chapter 7. When you use an LC circuit in an oscillator, the output signal of the oscillator will be at the resonance frequency of the LC circuit.

QUESTION

What is the formula for the oscillation (or resonant) frequency? _____

_____

ANSWER

$$f_r = \frac{1}{2\pi\sqrt{LC}}$$

In practice, the actual measured frequency is never quite the same as the calculated frequency. The capacitor and inductor values are not exact, and other stray capacitances in the circuit affect the frequency. When you need to set an exact frequency, use an adjustable capacitor or inductor.

12 Figure 9.9 shows the parallel LC circuit connected between the collector and the supply voltage, rather than between the collector and ground (as in Figure 9.8).

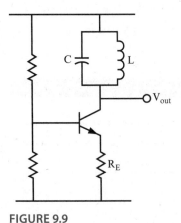

FIGURE 9.9

You can use this circuit and the circuit shown in Figure 9.8 to selectively amplify one frequency far more than others.

QUESTIONS

A. What would you expect this one frequency to be? _____

_____

B. Write the formula for the impedance of the circuit at the resonance frequency. ____

_____

C. What is the AC voltage gain at this frequency? _____

_____

ANSWERS

A. The resonance frequency

B. $Z = \dfrac{L}{C \times r}$

where r is the DC resistance of the coil

C. $A_V = \dfrac{Z}{R_E}$

13 Because of the low DC resistance of the coil, the DC voltage at the collector is usually close to the supply voltage (V_s). In addition, the AC output voltage positive peaks can exceed the DC level of the supply voltage. With large AC output, the positive peaks can actually reach $2V_s$, as shown in Figure 9.10.

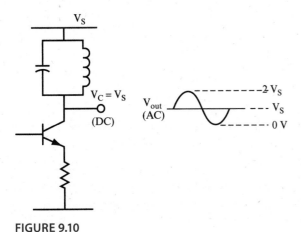

FIGURE 9.10

Indicate which of the following is an accurate description of the circuit in Figure 9.10:

A. Oscillator

B. Tuned amplifier

C. Common base circuit

D. Common emitter circuit

ANSWER

B. Tuned amplifier

FEEDBACK

14 To convert an amplifier into an oscillator, you must connect a portion of the output signal to the input. This feedback signal must be in phase with the input signal to induce oscillations.

Figure 9.11 shows three methods you can use to provide a feedback signal from a parallel LC circuit. Each is named for its inventor.

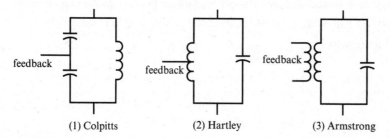

(1) Colpitts　　(2) Hartley　　(3) Armstrong

FIGURE 9.11

In the *Colpitts* method, the feedback signal is taken from a connection between two capacitors that form a voltage divider. In the *Hartley* method, the feedback signal is taken from a tap partway down the coil, or from a connection between two inductors. Therefore, an inductive voltage divider determines the feedback voltage. The *Armstrong* method uses a step-down transformer (an inductor with an extra coil with fewer turns than the main coil). In all three of these methods, between one-tenth and one-half of the output must be used as feedback.

QUESTIONS

A. Where is the feedback taken from in a Colpitts oscillator? _____

_____

B. What type of oscillator uses a tap on the coil for the feedback voltage? _____

_____

C. What type does not use a voltage divider? _____

_____

ANSWERS

A. A capacitive voltage divider

B. Hartley

C. Armstrong

15 The output voltage appears at one end of the parallel LC circuit shown in Figure 9.12, and the other end is effectively at ground. The feedback voltage V_f is taken between the junction of the two capacitors.

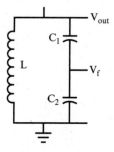

FIGURE 9.12

QUESTION

Using the voltage divider formula, what is V_f? _____

_____

ANSWER

$$V_f = \frac{V_{out} X_{C2}}{(X_{C1} + X_{C2})}$$

which becomes

$$V_f = \frac{V_{out} C_1}{(C_1 + C_2)}$$

16 To find the resonance frequency in this circuit, first find the equivalent total capacitance C_T of the two series capacitors. You then use C_T in the resonance frequency formula.

QUESTIONS

A. What is the formula for C_T? _____

_____

B. What is the resonance frequency formula for the Colpitts oscillator? _____

_____

ANSWERS

A. $C_T = \dfrac{C_1 C_2}{C_1 + C_2}$

B. $f_r = \dfrac{1}{2\pi\sqrt{LC_T}}$,

if Q is equal to or greater than 10

NOTE   If Q is less than 10, you can use one of the following two formulas to calculate the resonance frequency for a parallel LC circuit:

$$f_r = \frac{1}{2\pi\sqrt{LC}}\sqrt{1 - \frac{r^2 C}{L}} \quad \text{or} \quad f_r = \frac{1}{2\pi\sqrt{LC}}\sqrt{\frac{Q^2}{1 + Q^2}}$$

17 Figure 9.13 shows a parallel LC circuit in which the feedback voltage is taken from a tap N_1 turns from one end of a coil, and N_2 turns from the other end.

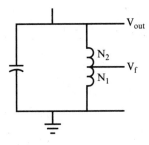

FIGURE 9.13

You can calculate the feedback voltage with a voltage divider formula that uses the number of turns in each part of the coil.

$$V_f = V_{out} \times \frac{N_1}{N_1 + N_2}$$

The manufacturer should specify N_1 and N_2.

QUESTIONS

A. Who invented this feedback method? _____

_____

B. When you divide V_f by $V_{out,}$ what is the result?_____

_____

ANSWERS

A. Hartley

B. Between one-tenth and one-half

18 Figure 9.14 shows a parallel LC circuit in which the feedback voltage is taken from the secondary coil of a transformer. The formula used to calculate the output voltage of a secondary coil is covered in problem 6 in Chapter 10, "The Transformer."

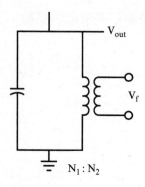

V_{out}

V_f

$N_1 : N_2$

FIGURE 9.14

QUESTION

Who invented this type of oscillator? _____

_____

ANSWER

Armstrong

19 For each of the feedback methods described in the last few problems, the voltage fed back from the output to the input is a fraction of the total output voltage ranging between one-tenth and one-half of V_{out}.

To ensure oscillations, the product of the feedback voltage and the amplifier voltage gain must be greater than 1.

$$A_V \times V_f > 1$$

It is usually easy to achieve this because A_v is much greater than 1.

No external input is applied to the oscillator. Its input is the small part of the output signal that is fed back. If this feedback is of the correct phase and amplitude, the oscillations start spontaneously and continue as long as power is supplied to the circuit.

The transistor amplifier amplifies the feedback signal to sustain the oscillations and converts the DC power from the battery or power supply into the AC power of the oscillations.

QUESTIONS

A. What makes an amplifier into an oscillator? _____

_____

B. What input does an amplifier need to become an oscillator? _____

_____

ANSWER

A. A resonant LC circuit with feedback of the correct phase and amount.

B. None. Oscillations happen spontaneously if the feedback is correct.

INSIDE THE INDUCTOR

When you use inductors, you should know how to deal with the different ways manufacturers label them.

Inductors are simply a coil of wire wrapped around a core, and some manufacturers leave them just like that. These inductors come with no markings, so you must keep them with the label from the packaging to identify them.

You can also find inductors that have a plastic coating around the wire coil. That coating is often marked with a numerical code that identifies the value of the inductor. The first two numbers are the first and second significant digits of the inductance value; the third number is the multiplier. (The units are µH, so an inductor marked with 101 has a value of 100 µH.)

Another method to mark inductors involves the same color code used for resistors. With this method, an inductor is marked with four color bands to show its value and tolerance. Some inductors have a wide silver band (about twice the width of the other bands) at the front of the color code bands. This wider band indicates that the component was built to a U.S. military specification and is not used to determine the inductance value.

Continued

(continued)

The value of each color used in the bands is shown in the following table (with units in µH):

Color	Significant Digits	Multiplier	Tolerance
Black	0	1	
Brown	1	10	±1 percent
Red	2	100	±2 percent
Orange	3	1,000	
Yellow	4	10,000	
Green	5		
Blue	6		
Violet	7		
Gray	8		
White	9		
Gold		0.1	±5 percent
Silver		0.01	±10 percent

The first two colored bands are the first and second significant digits of the inductance value. The third band is the multiplier and the fourth band represent the tolerance. For example, if an inductor is marked with blue, gray, red, and silver bands, its nominal inductance value is 6800 µH (6.8 mH) with a tolerance of ±10 percent .

Finally, you see inductors that have the value simply printed on them. This generally occurs on higher-value inductors, which are also physically larger.

Your best bet is to save the label on the packaging that the inductor comes in until you can check out the markings on the component.

THE COLPITTS OSCILLATOR

[20] Figure 9.15 shows a Colpitts oscillator circuit, the simplest of the LC oscillators to build.

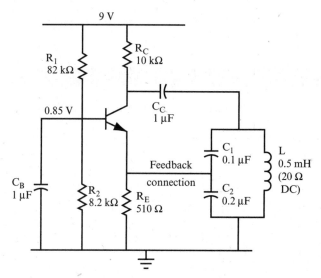

FIGURE 9.15

The feedback signal is taken from the capacitive voltage divider and fed to the emitter. This connection provides a feedback signal to the emitter in the phase required to provide positive feedback.

In this circuit, the reactance of capacitor C_B is low enough for the AC signal to pass through it, rather than passing through R_2. Capacitor C_B should have a reactance, X_{CB}, of less than 160 ohms at the oscillation frequency. If R_2 happens to be smaller than 1.6 kΩ, choose a value of X_{CB} that is less than one-tenth of R2.

QUESTION

For the circuit shown in Figure 9.15, what is your first estimate for C_B? Assume that f_r is equal to 1 kHz, and that X_{CB} equals 160 ohms. _____

_____

ANSWER

$$X_{CB} = 160\,\text{ohms} = \frac{1}{2\pi f C_B} = \frac{1}{2\times\pi\times10^3\times C_B}$$

Therefore, $C_B = 1\ \mu F$; larger values of C_B also work.

21 Use the Colpitts oscillator component values shown in Figure 9.15 to answer the following questions.

QUESTIONS

A. What is the effective total capacitance of the two series capacitors in the tuned circuit?

$C_T =$ _____

B. What is the oscillator frequency?

$f_r =$ _____

C. What is the impedance of the tuned circuit at this frequency?

$Z =$ _____

D. What fraction of the output voltage is fed back?

$V_f =$ _____

E. What is the reactance of C_B at the frequency of oscillation?

$X_{CB} =$ _____

ANSWERS

A. $C_T = 0.067 \ \mu F$.

B. Because Q is not known, use the formula that includes the resistance of the coil (see problem 16):

$f_r = 26.75$ kHz.

If you use the calculated value of f_r to calculate Q, as in problem 20 of Chapter 7, you find that $Q = 4.2$. Therefore, it is appropriate to use the formula that includes the resistance of the coil to calculate f_r.

C. Use the following:

$$Z_T = \frac{L}{rC}, Z_T = 373 \text{ ohms}.$$

D. Use a voltage divider with the capacitor values:

$$V_f = V_{out} \frac{C_1}{(C_1 + C_2)} = \frac{V_{out}}{3}.$$

E. $X_{CB} =$ about 6 ohms, which is a good value (much less than the 8200 ohm value of R_2).

22 Figure 9.16 shows a Colpitts oscillator circuit that uses a different method for making feedback connections between the parallel LC circuit and the transistor.

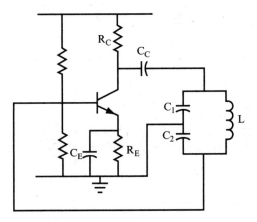

FIGURE 9.16

List the differences between this circuit and the one shown in Figure 9.15. _____

_____

The feedback is connected to the base instead of the emitter, and the ground is connected to the center of the capacitive voltage divider. The capacitor C_E has been added. (This connection provides a feedback signal to the base in the correct phase to provide positive feedback.)

23 In the circuit shown in Figure 9.16, capacitor C_E should have a reactance of less than 160 ohms at the oscillation frequency. If the emitter resistor R_E is smaller in value than 1.6 kΩ, then C_E should have a reactance that is less than $R_E/10$ at the oscillation frequency.

If you use an emitter resistor of 510 ohms in a 1 kHz oscillator, what value of capacitor should you use for C_E? _____

_____

ANSWER

$$X_C = \frac{510}{10} = \frac{1}{2\pi f C_E} = \frac{0.16}{10^3 \times C_E}$$

So, $C_E = 3.2\ \mu F$. Thus, you should use a capacitor larger than 3 μF.

PROJECT 9.1: The Colpitts Oscillator

OBJECTIVE

The objective of this project is to demonstrate that an oscillator generates a sine wave when feedback is applied to either the emitter or base.

GENERAL INSTRUCTIONS

When the Colpitts oscillator circuit with feedback to the emitter is set up, you use your oscilloscope to measure the period of the waveform. Then you change the circuit to provide feedback to the base, and again use your oscilloscope to measure the period of the waveform. This data enables you to calculate the frequency of the sine wave generated in each case.

Parts List

You need the following equipment and supplies:

- ❏ One 10 kΩ, 0.25-watt resistor.
- ❏ One 510 Ω, 0.25-watt resistor.
- ❏ One 82 kΩ, 0.25-watt resistor.
- ❏ One 8.2 kΩ, 0.25-watt resistor.
- ❏ Two 1 µF capacitors (This value of capacitor is available in either polarized or unpolarized versions. You should get unpolarized capacitors for this application.)
- ❏ One 0.1 µF capacitor.
- ❏ One 0.22 µF capacitor.
- ❏ One 4.7 µF capacitor. (This value of capacitor is usually polarized, which is fine for this position in the circuit.)

❑ One 0.5 mH inductor. (Suppliers may also refer to this value as 500 μH.)

❑ One 9-volt battery pack.

❑ One breadboard.

❑ One oscilloscope.

❑ One PN2222 transistor. Figure 9.17 shows the pinout diagram for PN2222 transistors.

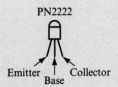

FIGURE 9.17

STEP-BY-STEP INSTRUCTIONS

Set up Circuit #1, the Colpitts oscillator circuit with feedback to the emitter, as shown in Figure 9.18. If you have some experience in building circuits, this schematic (along with the previous parts list) should provide all the information you need to build the circuit. If you need a bit more help building the circuit, look at the photos of the completed circuit in the "Expected Results" section.

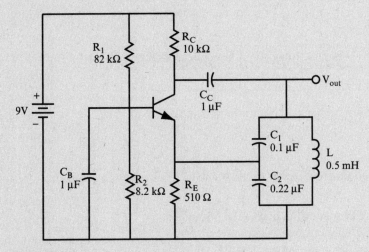

FIGURE 9.18

Carefully check your circuit against the diagram.

After you have checked your circuit, follow these steps and record your measurements in the blank table following the steps.

1. Connect the oscilloscope probe for Channel 1 to a jumper wire connected to V_{out}. Connect the ground clip to a jumper wire attached to the ground bus.

2. Measure and record the period of the sine wave.

3. Disconnect the battery from the circuit, and make the changes required to set up Circuit #2, the Colpitts oscillator circuit with feedback to the base, as shown in Figure 9.19.

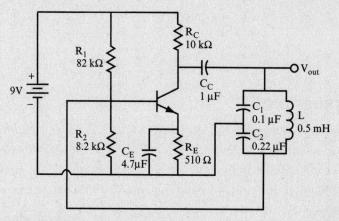

FIGURE 9.19

4. Repeat steps 1 and 2.

Circuit #	Period (μsec)	Frequency (kHz)
1		
2		

EXPECTED RESULTS

Figure 9.20 shows the breadboarded Colpitts oscillator with feedback to the emitter (Circuit# 1).

Figure 9.21 shows an oscilloscope attached to the circuit.

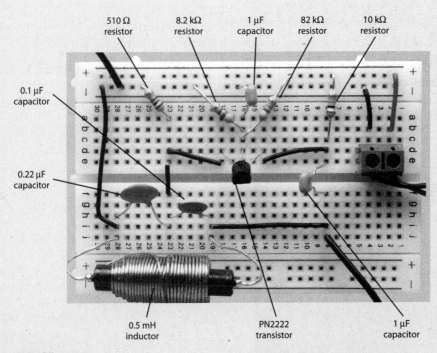

510 Ω resistor

8.2 kΩ resistor

1 µF capacitor

82 kΩ resistor

10 kΩ resistor

0.1 µF capacitor

0.22 µF capacitor

0.5 mH inductor

PN2222 transistor

1 µF capacitor

FIGURE 9.20

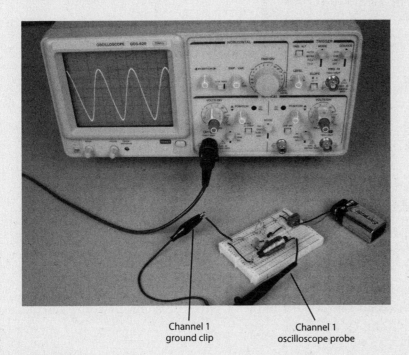

Channel 1 ground clip

Channel 1 oscilloscope probe

FIGURE 9.21

Figure 9.22 shows the sine wave generated by the Colpitts oscillator with feedback to the emitter. You can determine the period of this waveform by counting the number of horizontal divisions the waveform takes to complete one cycle, and then multiplying the number of divisions by the TIME/DIV setting.

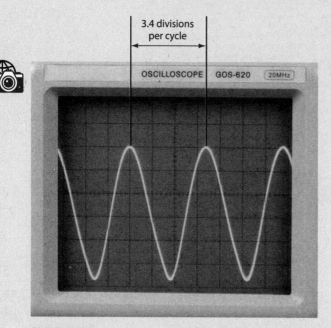

FIGURE 9.22

As you measure the period, you may need to adjust the TIME/DIV, the horizontal POSITION, and the vertical POSITION controls on the oscilloscope. The controls shown in Figure 9.23 are adjusted to measure the period for the Colpitts oscillator.

Figure 9.24 shows the breadboarded Colpitts oscillator with feedback to the base (Circuit# 2).

Figure 9.25 shows the sine wave generated by the Colpitts oscillator with feedback to the base. You can determine the period of this waveform by counting the number of horizontal divisions the waveform takes to complete one cycle, and then multiplying the number of divisions by the TIME/DIV setting.

The oscilloscope connections and oscilloscope control panel settings for the Colpitts oscillator with feedback to the base are not shown. They are the same as the oscilloscope connections and oscilloscope control panel for the Colpitts oscillator with feedback to the emitter.

Horizontal
position knob

Time/div control set
to 10 μsec/div

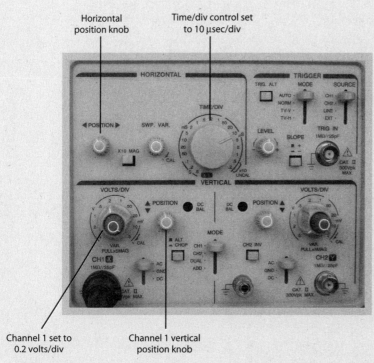

Channel 1 set to
0.2 volts/div

Channel 1 vertical
position knob

FIGURE 9.23

Feedback
to base

4.7 μF
capacitor

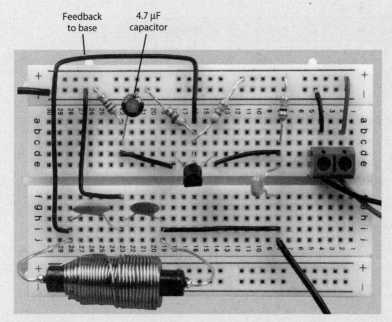

FIGURE 9.24

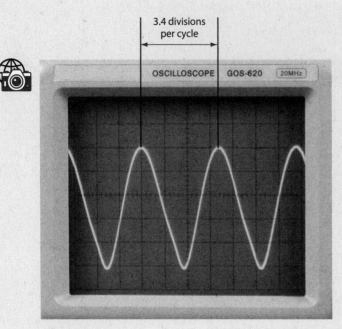

3.4 divisions per cycle

OSCILLOSCOPE   GOS-620   20MHz

FIGURE 9.25

Your values should be close to those shown in the following table:

Circuit #	Period (μsec)	Frequency (kHz)
1	34	29.4
2	34	29.4

Notice that the frequency of the sine wave generated by both circuits is the same. This demonstrates that an oscillator can function with feedback to either the emitter or base of the transistor.

24 Figure 9.26 shows a Colpitts oscillator with the parallel LC circuit connected between the collector and the supply voltage. As with the circuits shown in Figures 9.15 and 9.16, this circuit provides a feedback signal to the transistor (in this case, the emitter) in the correct phase to provide positive feedback.

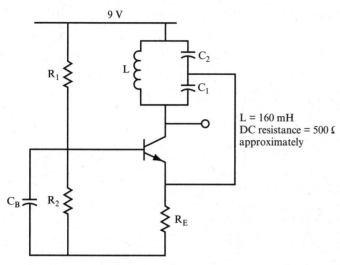

FIGURE 9.26

The following table shows possible values you might use for C_1 and C_2 in the circuit shown in Figure 9.26:

C_1	C_2	C_T	f_r
0.01 μF	0.1 μF		
0.01 μF	0.2 μF		
0.01 μF	0.3 μF		
0.1 μF	1 μF		
0.2 μF	1 μF		

QUESTIONS

A. Calculate C_T and f_r for each row of the preceding table. _____

_____

B. Does increasing C_2, while holding C_1 constant, increase or decrease the resonance frequency? _____

_____

C. What effect does increasing C_1 have on the resonance frequency? _____

_____

D. What is the condition that results in the highest possible resonance frequency? ___

_____

E. What would be the highest resonance frequency if C_1 is fixed at 0.01 µF, and C_2 can vary from 0.005 µF to 0.5 µF? _____

_____

ANSWERS

A. The following table shows the values of C_T and f_r:

C_1	C_2	C_T	f_r
0.01 µF	0.1 µF	0.009 µF	4.19 kHz
0.01 µF	0.2 µF	0.0095 µF	4.08 kHz
0.01 µF	0.3 µF	0.0097 µF	4.04 kHz
0.1 µF	1 µF	0.09 µF	1.33 kHz
0.2 µF	1 µF	0.167 µF	0.97 kHz

B. Increasing C_2 decreases the resonance frequency, and, therefore, decreases the output frequency of the oscillator.

C. Increasing C_1 also decreases the resonance frequency and the output frequency of the oscillator.

D. When C_T is at its lowest possible value.

E. When C_2 is 0.005 µF, C_T will be 0.0033 µF, which is its lowest possible value. Therefore, the frequency is at the highest possible value, or approximately 6.9 kHz. The lowest frequency occurs when C_2 is at its highest setting of 0.5 µF.

THE HARTLEY OSCILLATOR

25 Figure 9.27 shows a Hartley oscillator circuit. In this type of circuit, the feedback is taken from a tap on the coil, or from a connection between two inductors.

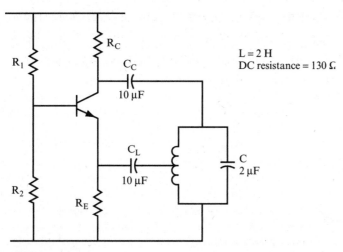

$L = 2$ H
DC resistance = 130 Ω

FIGURE 9.27

Capacitor C_L stops the emitter DC voltage from being pulled down to 0 volts through the coil. C_L should have a reactance of less than $R_E/10$, or less than 160 ohms at the oscillator frequency.

QUESTIONS

Work through the following calculations:

A. What is the resonance frequency?

$f_r =$ _____

B. What is the approximate impedance of the load?

$Z =$ _____

C. What missing information prevents you from calculating the fraction of the voltage drop across the coil that is fed back to the emitter?

_____

ANSWERS

A. 80 Hz (approximately).

B. 7.7 kΩ (approximately).

C. The number of turns in the coil and the position of the tap are not known.

Figure 9.28 shows a Hartley oscillator with the parallel LC circuit connected between the collector and the supply voltage. As with the circuit shown in Figure 9.27, this circuit provides a feedback signal to the emitter from a tap in the coil, in the correct phase to provide positive feedback.

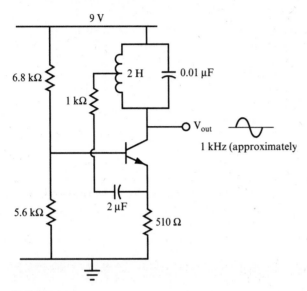

FIGURE 9.28

PROJECT 9.2: The Hartley Oscillator

OBJECTIVE

The objective of this project is to demonstrate a Hartley oscillator using two inductors in series.

GENERAL INSTRUCTIONS

After the Hartley oscillator circuit is set up, you use your oscilloscope to measure the period of the waveform, from which you can calculate the frequency of the oscillator. You also calculate the frequency from the inductance and capacitance used in the parallel LC circuit. Note that when two inductors in series are used, rather than a tapped coil, the total inductance is found by adding the individual inductance values, using the following equation:

$$L_T = L_1 + L_2$$

Parts List

You need the following equipment and supplies:

- ❏ One 10 kΩ, 0.25-watt resistor.
- ❏ One 510 Ω, 0.25-watt resistor.
- ❏ One 82 kΩ, 0.25-watt resistor.
- ❏ One 8.2 kΩ, 0.25-watt resistor.
- ❏ Three 1 μF capacitors. (This value of capacitor is available in either polarizied or unpolarized versions.You should get unpolarized capacitors for this application.)
- ❏ One 0.01 μF capacitor.
- ❏ One 6.8 mH inductor.
- ❏ One 3.1 mH inductor.
- ❏ One 9-volt battery pack.
- ❏ One breadboard.
- ❏ One oscilloscope.
- ❏ One PN2222 transistor. Figure 9.29 shows the pinout diagram for PN2222 transistors.

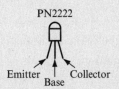

Emitter Base Collector

FIGURE 9.29

STEP-BY-STEP INSTRUCTIONS

Set up the Hartley oscillator circuit shown in Figure 9.30. If you have some experience in building circuits, this schematic (along with the previous parts list) should provide all the information you need to build the circuit. If you need a bit more help building the circuit, look at the photos of the completed circuit in the "Expected Results" section.

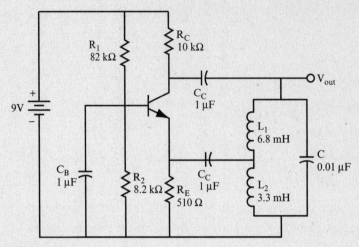

FIGURE 9.30

Carefully check your circuit against the diagram.

When you have checked your circuit, follow these steps.

1. Connect the oscilloscope probe for Channel 1 to a jumper wire connected to V_{out}. Connect the ground clip to a jumper wire attached to the ground bus.

2. Measure and record the period of the sine wave.

    Period = _____

3. Calculate the frequency of the sine wave.

    Frequency = _____

4. Calculate the expected resonance frequency from the value of the capacitor and inductors used in the parallel LC circuit using the following equation:

$$f_r = \frac{1}{2\pi\sqrt{L_T C}}$$

    f_r = _____

EXPECTED RESULTS

Figure 9.31 shows the breadboarded Hartley oscillator.

Figure 9.32 shows an oscilloscope attached to the circuit.

510 Ω
resistor

8.2 kΩ
resistor

1 µF
capacitor

82 kΩ
resistor

10 kΩ
resistor

1 µF
capacitor

3.3 mH
inductor

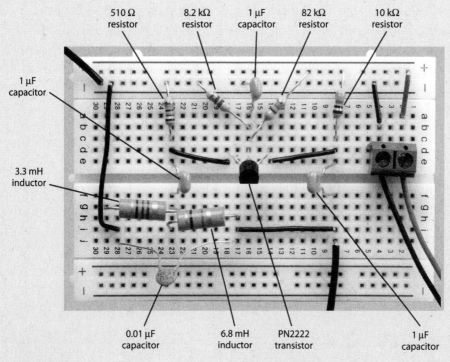

0.01 µF
capacitor

6.8 mH
inductor

PN2222
transistor

1 µF
capacitor

FIGURE 9.31

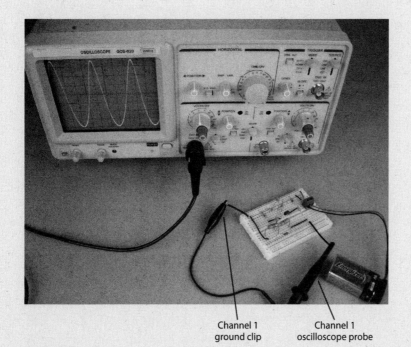

Channel 1
ground clip

Channel 1
oscilloscope probe

FIGURE 9.32

Figure 9.33 shows the sine wave generated by the Hartley oscillator. You can determine the period of this waveform by counting the number of horizontal divisions the waveform takes to complete one cycle, and then multiplying the number of divisions by the TIME/DIV setting.

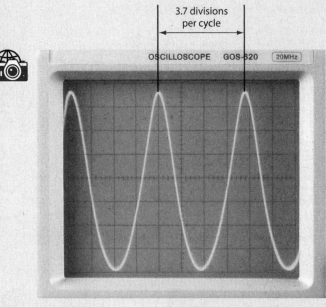

FIGURE 9.33

As you measure the period, you may need to adjust the TIME/DIV, the horizontal POSITION, and the vertical POSITION controls on the oscilloscope. The controls shown in Figure 9.34 are adjusted to measure the period for the Hartley oscillator.

Your values should be close to those shown here:

■ Period = 74 μsec

■ Frequency = 13.5 kHz

This measured frequency is close to the calculated resonance frequency of 15.8 kHz.

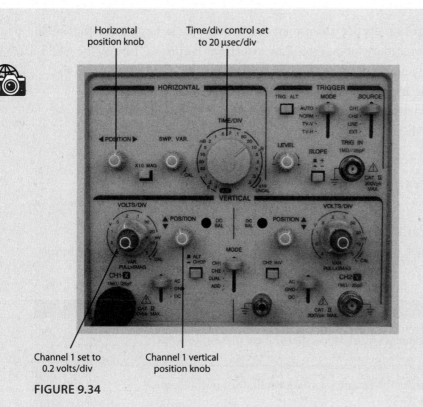

Horizontal
position knob

Time/div control set
to 20 μsec/div

Channel 1 set to
0.2 volts/div

Channel 1 vertical
position knob

FIGURE 9.34

THE ARMSTRONG OSCILLATOR

The Armstrong oscillator shown in Figure 9.35 is somewhat more difficult to design and build. Here, the oscillations depend more on the extra winding on the coil than on any other factor.

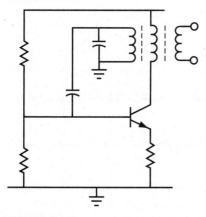

FIGURE 9.35

Because of the large variety of transformers and coils available, it is almost impossible to give you a simple procedure for designing an Armstrong oscillator. Instead, the manufacturer specifies the number of turns required on the coils, which guarantees that the oscillator will work in its most common operation, at high radio frequencies.

Because of the practical difficulties, the Armstrong oscillator and its variations are not explored any further.

PRACTICAL OSCILLATOR DESIGN

26 This section briefly covers some practical problems with oscillators.

Before you proceed, review the important points of this chapter by answering the following questions.

QUESTIONS

A. What three elements must an oscillator have present to work? _____

_____

B. What determines the frequency of an oscillator's output signal?_____

_____

C. What provides the feedback?_____

_____

D. How many feedback methods for oscillators have been discussed?_____

_____

E. What do you need to start the oscillations once the circuit has been built?_____

_____

ANSWERS

A. An amplifier, a resonant LC circuit (or some other frequency determining components), and feedback.

B. The frequency of the output signal is the same as the resonance frequency.

C. A voltage divider on the resonant circuit.

D. Three: the Colpitts, Hartley, and the Armstrong.

E. Nothing: The oscillations should start spontaneously if the component values in the circuit are correct.

The main practical problem with building oscillators is selecting the coil. For mass production, a manufacturer can specify and purchase the exact coil required. But in a lab or workshop (where you are building only a single circuit), it is often difficult or impossible to find the exact inductor specified in a circuit design. What usually happens is that you use the most readily available coil, and design the rest of the circuit around it. This presents three possible problems:

- You may not know the exact value of the inductance.
- The inductance value may not be the best for the wanted frequency range.
- The coil may or may not have tap points or extra windings, and this may cause a change in the circuit design. For example, if there are no taps, then you cannot build a Hartley oscillator.

Because Colpitts is the easiest oscillator to make work in practice, and provides an easy way around some of the practical difficulties, you can focus on that oscillator.

You can use almost any coil when building a Colpitts oscillator, provided it is suitable for the frequency range you want. For example, a coil from the tuner section of a television set would not be suitable for a 1-kHz audio oscillator because its inductance value is outside the range best suited to a low-frequency audio circuit.

SIMPLE OSCILLATOR DESIGN PROCEDURE

27 Following is a simple step-by-step procedure for the design of a Colpitts oscillator. The Colpitts can work over a wide frequency range. (A Hartley can be designed using a similar set of steps.)

By following this procedure, you can design an oscillator that works in the majority of cases. There is a procedure you can use that guarantees that the oscillator will work, but it is far more complex.

Follow these steps:

1. Choose the frequency of the oscillator output signal.

2. Choose a suitable coil. This step presents the greatest practical difficulty. Some values of coil are often not available, so you must use whatever is readily available. Fortunately, you can use a wide range of inductance values and still obtain the desired resonance frequency by adjusting the value of the capacitor.

3. If you know the value of the inductance, calculate the capacitor value using this formula:

$$f_r = \frac{1}{2\pi\sqrt{LC}}$$

Use this value of capacitor for C_1 in the next steps.

4. If you don't know the inductance value, choose any value of capacitance and call this C_1. This may produce a frequency considerably different from what you require. However, at this stage, the main thing is to get the circuit oscillating. You can adjust values later.

5. Choose a capacitor C_2 that is between 3 and 10 times the value of C_1. Figure 9.36 shows the two capacitors and the coil connected in a parallel circuit, with the two capacitors acting as a voltage divider.

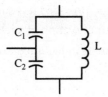

FIGURE 9.36

At this point, stop and make some assumptions. Suppose you need a frequency of 10 kHz and have a coil with a 16 mH inductance.

QUESTIONS

A. What approximate value of C_1 do you need? _____

_____

B. What value of C_2 do you need? _____

_____

ANSWERS

A. $C_1 = 0.016\ \mu F$

B. $C_2 = 0.048\ \mu F$ to $0.16\ \mu F$

28 Now, continue with the design procedure by following the next steps.

6. Design an amplifier with a common emitter gain of about 20. Choose a collector DC voltage that is about half the supply voltage. The main point to keep in mind here is that the collector resistor R_C should be about one-tenth the value of the impedance of the LC circuit at the resonance frequency. This is often a difficult choice to make, especially if you don't know the coil value. Usually, you have to make an assumption, so R_C is an arbitrary choice.

7. Draw the circuit.

8. Calculate the value of C_C. Do this by making X_C 160 ohms at the desired frequency. This is another "rule of thumb" that happens to work, and you can justify it mathematically. Use the following formula:

$$C_C = \frac{1}{2\pi f_r X_C}$$

QUESTION

Substitute the values given so far into the formula to calculate C_C. _____

_____

ANSWER

$$C_C = \frac{1}{2\pi \times 10\,\text{kHz} \times 160\,\Omega} = 0.1\,\mu F$$

29 Now, complete one last step.

9. Calculate the value of C_B. Again, choose a value so that X_C is 160 ohms at the desired frequency.

QUESTION

What is the value of C_B? _____

_____

ANSWER

$C_B = 0.1\,\mu F$

---

30 Continue the design procedure steps.

10. After you build an oscillator, apply power to the circuit and look at the output signal on an oscilloscope. If the output signal is oscillating, check the frequency. If the frequency varies significantly from the desired frequency, then change C_1 until you get the wanted frequency. Change C_2 to keep the ratio of the capacitance values about the same as discussed in step 5. C_2 affects the output level.

11. If the circuit does not oscillate, go through the steps outlined in the troubleshooting checklist that follows.

OSCILLATOR TROUBLESHOOTING CHECKLIST

If an oscillator does not work, most often the trouble is with the feedback connections. A little experimenting (as outlined in steps 2 through 6 of the following checklist) should produce the right results. This is especially true when you use an unknown coil that may have several taps or windings. However, you should try each of the following steps if you have trouble.

1. Ensure that C_B, C_C, and C_E are all large enough to have a reactance value less than 160 ohms. Ensure that C_E is less than one-tenth of R_E.

2. Check the C_1/C_2 ratio. It should be between 3:1 and 10:1.

3. Swap out C_1 and C_2. They may be connected to the wrong end of the LC circuit.

4. Check that you made the feedback connection to and from the correct place.

5. Check both ends of the LC circuit to see that they are connected to the correct place.

6. Check the DC voltage level of the collector, base, and emitter.

7. Check the capacitor values of the LC circuit. If necessary, try some other values until the circuit oscillates.

8. If none of the previous actions produce oscillations, check to see if any of the components are defective. The coil may be opened or shorted. The capacitor may be shorted. The transistor may be dead, or its β may be too low. Check the circuit wiring carefully.

In most cases, one or more of these steps produces oscillations.

When an oscillator works, it may still have one or two main faults, including the following:

- *Distorted output waveform*—This can happen when C_B, C_C, or C_E are not low enough in value, or when an output amplitude is too high.

- *Output level too low*—When this happens, the sine wave is usually "clean" and "pure." In a Colpitts oscillator, changing the ratio of C_1 and C_2 often helps raise the output level. If not, you can use another transistor as an amplifier after the oscillator, as discussed in Chapter 8, problem 21.

31 Now, work through a design example. Design an oscillator with an output frequency of 25 kHz using a coil with a value of 4 mH, and address each of the steps in problems 27–30 as described in these questions.

QUESTIONS

1. The value of f_r is given as 25 kHz.

2. L is given as 4 mH.

3. Use the formula to find C_1.

   $C_1 =$ _____

4. You do not need this step.

5. Choose C_2.

   $C_2 =$ _____

6. The procedure to design amplifiers is shown in Chapter 8.

7. The circuit is shown in Figure 9.37.

8. Find C_C.

   $C_C =$ _____

9. Find C_B.

   $C_B =$ _____

ANSWERS

$C_1 = 0.01 \ \mu F$

$C_2 = 0.1 \ \mu F$

$C_C = 0.047 \ \mu F$ (use $0.1 \ \mu F$)

$C_B = 0.047 \ \mu F$ (use $0.1 \ \mu F$)

Steps 10–11 are the procedure you use to ensure that the oscillator works. If you built this circuit, go through steps 10–11. You don't need to do them if you didn't actually build the circuit.

---

32 Figure 9.37 shows the circuit designed in problem 31.

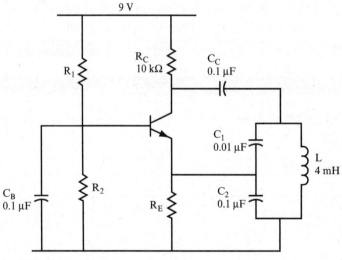

FIGURE 9.37

Measurements of the output signal of this oscillator confirm a frequency close to 25 kHz.

QUESTION

Find the impedance of the LC circuit at resonance. Note that r (the DC resistance of the inductor) is 12 ohms. _____

_____

ANSWER

$$Z = \frac{L}{C \times r} = \frac{4 \times 10^{-3}}{0.01 \times 10^{-6} \times 12} = 33\,k\Omega \text{ (approximately)}$$

This is about three times the value used for R_C, instead of being 10 times the value of R_C, as suggested in step 6 of problem 28.

33 If you want, work through this second oscillator design example. Design an oscillator with an output frequency of 250 kHz using a coil with a value of 500 μH.

QUESTIONS

1. $f_r = 250$ kHz

2. $L = 500$ μH $= 0.5$ mH

3. Find C_1.

   $C_1 =$ _____

4. You do not need this step.

5. Find C_2.

   $C_2 =$ _____

6. Use the same amplifier as in the last example.

7. The circuit is shown in Figure 9.38.

8. Find C_C.

$C_C =$ _____

9. Find C_B.

$C_B =$ _____

ANSWER

$C_1 = 0.0008\ \mu F$; therefore, choose a standard value of $0.001\ \mu F$.

$C_2 = 0.0047\ \mu F$, which is a standard value.

$C_B = C_C = 0.004\ \mu F$ (minimum).

34 The circuit you designed in problem 33 is shown in Figure 9.38.

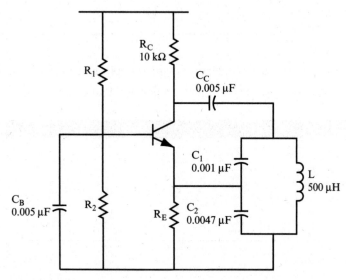

FIGURE 9.38

Measurements of the output signal of this oscillator confirm a frequency close to 250 kHz.

QUESTION

Find the impedance of the LC circuit at resonance. Note that r (the DC resistance of the inductor) is 20 ohms. _____

_____

ANSWER

$Z = 30\ \text{k}\Omega$

This is about 3 times the value of R_C, rather than 10 times the value of R_C, as suggested in step 6 of problem 28.

35 Figure 9.39 shows several other oscillator circuits. Calculate the expected output frequency for each circuit and build as many as you want. Check the measured oscillator output frequency against the calculated values for each circuit you build.

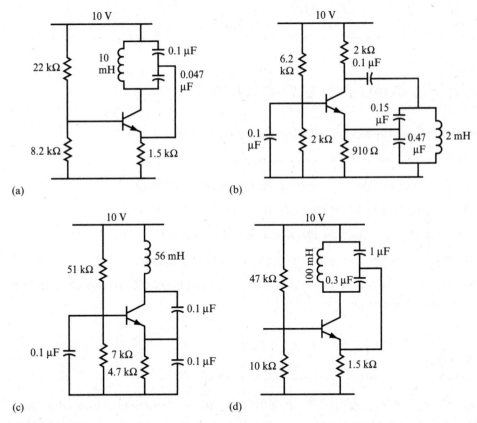

FIGURE 9.39

What is the output frequency for each circuit?

A. f = _____

B. $f =$ _____

C. $f =$ _____

D. $f =$ _____

ANSWER

A. 8.8 kHz

B. 10 kHz

C. 3 kHz

D. 1 kHz

SUMMARY AND APPLICATIONS

This chapter covered the following topics related to oscillators:

- The main elements that make up an oscillator
- How to differentiate between positive and negative feedback
- The type of feedback that causes a circuit to oscillate
- Two methods to obtain feedback in an oscillator circuit
- How resonant LC circuits set the frequency of an oscillator

You also practiced designing a simple oscillator circuit to solidify your understanding of its elements and operation.

SELF-TEST

These questions test your understanding of the concepts and equations presented in this chapter. Use a separate sheet of paper for your diagrams or calculations. Compare your answers with the answers provided following the test.

1. What are the three sections necessary in an oscillator? _____

_____

2. What is the difference between positive and negative feedback?_____

_____

3. What type of feedback is required in an oscillator?_____

_____

4. What is the formula for the frequency of an oscillator?_____

_____

5. Draw the circuit for a Colpitts oscillator.

6. Draw the circuit for a Hartley oscillator.

7. Draw the circuit for an Armstrong oscillator.

8. Problems 27–30 give a design procedure for oscillators. How well do the circuits in problem 35 fulfill the criteria for that procedure? In other words, check the values of V_p, A_V (for a common emitter amplifier), C_1/C_2 ratio, R_C/Z ratio, and the frequency.

   A. _____

   B. _____

   C. _____

   D. _____

9. For the circuit shown in Figure 9.38, calculate the values of C_1, C_2, C_C, and C_B for an oscillator with an output frequency of 10 kHz using a 100 mH coil.

ANSWERS TO SELF-TEST

If your answers do not agree with those provided here, review the problems indicated in parentheses before you go on to Chapter 10.

1.	An amplifier, feedback, and a resonant load.	(problem 1)
2.	Positive feedback is "in phase" with the input, and negative feedback is "out of phase" with the input.	(problems 2–3)
3.	Positive feedback.	(problem 3)
4.	$f_r = \dfrac{1}{2\pi\sqrt{LC}}$	(problem 11)
5.	See Figure 9.15.	(problem 20)

Continued

(continued)

6.	See Figure 9.27.	(problem 25)
7.	See Figure 9.35.	(problem 25)
8A.	$V_f = \dfrac{0.047}{0.147}$	(problems 27–30)
	A_V cannot be calculated.	
	$C_1/C_2 = 0.047/0.1 = 0.47$	
	Z cannot be calculated because r is unknown.	
	$f_r = 8.8$ kHz (approximately).	
8B.	$V_f = \dfrac{0.15}{0.62}$	
	$A_V = 2.2$ (approximately).	
	$C_1/C_2 = 1/3$ (approximately).	
	Z cannot be calculated.	
	$f_r = 10$ kHz (approximately).	
8C.	$V_f = \dfrac{0.1}{0.2}$	
	A_V cannot be calculated.	
	$C_1/C_2 = 1$	
	Z cannot be calculated.	
	$f_r = 3$ kHz	
8D.	$V_f = \dfrac{0.3}{1}$	
	A_V cannot be calculated.	
	$C_1/C_2 = 0.3$	
	Z cannot be calculated.	
	$f_r = 1$ kHz (approximately)	
9.	$C_1 = 0.0033$ μF; $C_2 = 0.01$ μF; $C_B = C_C = 0.1$ μF	(problems 26–30)

10

The Transformer

Transformers are used to "transform" an AC voltage to a higher or lower level. When you charge your cellphone, you use a transformer to reduce the 120 volts supplied by the wall outlet to the 5 volts or so needed to charge your cellphone's battery. Most electrical devices that you plug into wall outlets use transformers to reduce power coming from an outlet to that required by the electrical components in the device.

You can also use transformers to increase voltage. For example, some of the equipment used to manufacture integrated circuits requires thousands of volts to operate. Transformers are used to increase the 240 volts supplied by the power company to the required voltage.

When you complete this chapter, you will be able to do the following:

- Recognize a transformer in a circuit.

- Explain and correctly apply the concepts of *turns ratio* and *impedance matching*.

- Recognize two types of transformer.

- Do simple calculations involving transformers.

TRANSFORMER BASICS

1 Consider two coils placed close to each other, as shown in Figure 10.1. If you apply an AC voltage to the first (or primary) coil, the alternating current flowing through the coil creates a fluctuating magnetic field that surrounds the coil. As the strength and polarity of this magnetic field changes, it induces an AC voltage in the second (or secondary) coil. The AC signal induced in the secondary coil is at the same frequency as the AC signal applied to the primary coil.

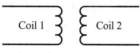

FIGURE 10.1

Both transformer coils are usually wound around a core made of a magnetic material such as iron or ferrite to increase the strength of the magnetic field.

QUESTIONS

A. When the two coils are wound around the same core, are they connected electrically? _____

_____

B. What type of device consists of two wire coils wound around an iron or ferrite core? _____

_____

C. If you apply an AC voltage to the terminals of the primary coil, what occurs in the secondary coil? _____

_____

ANSWERS

A. No.

B. A transformer.

C. An AC voltage of the same frequency is induced in the secondary coil.

---

2 A transformer is used only with alternating currents. A fluctuating magnetic field (such as that generated by alternating current flowing through a primary coil) is required to induce current in a secondary coil. The stationary magnetic field generated by direct current flowing through a primary coil will not induce any current or voltage in a secondary coil.

When a sine wave signal is applied to a primary coil, you can observe a sine wave of the same frequency across the secondary coil, as shown in Figure 10.2.

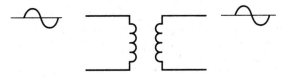

FIGURE 10.2

QUESTIONS

A. What will be the difference in frequency between a signal applied to a primary coil and the signal induced in a secondary coil? _____

_____

B. What will be the voltage difference across a secondary coil if 10 volts DC is applied to the primary coil? _____

_____

ANSWERS

A. No difference. The frequencies will be the same.

B. Zero volts. When a DC voltage is applied to the primary coil, there is no voltage or current induced in the secondary coil. You can summarize this by saying that DC does not pass through a transformer.

3 You can compare the output waveform measured between the terminals of the secondary coil to the output waveform measured between the terminals of the primary coil. If the output goes positive when the input goes positive, as shown in Figure 10.3, then they are said to be *in phase*.

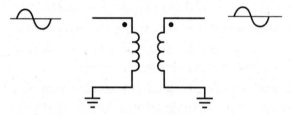

FIGURE 10.3

The dots on the coils in Figure 10.3 indicate the corresponding end of each coil. If one coil is reversed, then the output will be inverted from the input. The output is said to be *out of phase* with the input, and a dot is placed at the opposite end of the coil.

QUESTION

In Figure 10.4, the output sine wave is out of phase with the input sine wave. Place a dot in the correct location in the secondary coil to show that it is out of phase.

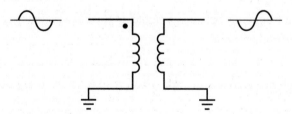

FIGURE 10.4

4 The transformer shown in the right side of Figure 10.5 has three terminals. The additional terminal, in the middle of the coil, is called a *center tap*.

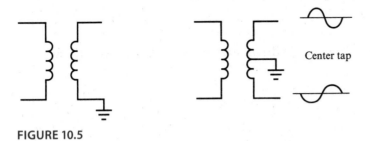

FIGURE 10.5

QUESTION

What is the difference between the two output waveforms shown for the transformer on the right side of Figure 10.5? _____

_____

ANSWER

The two waveforms are 180 degrees out of phase. That is, the positive peak of the upper output occurs at the same time as the negative peak of the lower waveform.

5 In a transformer, the output voltage from the secondary coil is directly proportional to the number of turns of wire in the secondary coil. If you increase the number of turns of wire in the secondary coil, a larger output voltage is induced across the secondary coil. If you decrease the number of turns of wire in the secondary coil, a smaller output voltage is induced across the secondary coil.

QUESTION

How does increasing the number of turns of wire in a secondary coil affect the output voltage across the secondary coil? _____

_____

ANSWER

It increases the output voltage across the secondary coil.

6 Figure 10.6 shows the number of turns in the primary and secondary coils as N_p and N_s.

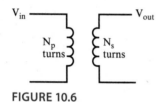

FIGURE 10.6

QUESTION

The ratio of the input to output voltage is the same as the ratio of the number of turns in the primary coil to the number of turns in the secondary coil. Write a simple formula to express this. _____

_____

ANSWER

$$\frac{V_{in}}{V_{out}} = \frac{N_p}{N_s}$$

NOTE The ratio of primary turns to secondary turns is called the *turns ratio (TR)*:

$$TR = \frac{N_p}{N_s} = \frac{V_{in}}{V_{out}}$$

7 Use the formula from problem 6 to answer the following question.

QUESTION

Calculate the output voltage of a transformer with a 2 to 1 (2:1) turns ratio when you apply a 10 V_{pp} sine wave to the primary coil. _____

_____

ANSWER

$$\frac{V_{in}}{V_{out}} = \frac{N_p}{N_s} = TR$$

$$V_{out} = V_{in}\frac{N_s}{N_p} = V_{in} \times \frac{1}{TR}$$

$$V_{out} = V_{in} \times \frac{1}{TR} = 10 \times \frac{1}{2} = 5\ V_{pp}$$

8 Use the input voltage and turns ratio for a transformer to answer the following questions.

QUESTIONS

Calculate V_{out} in the following:

A. $V_{in} = 20\ V_{pp}$, turns ratio = 5:1.

$V_{out} =$ _____

B. $V_{in} = 1\ V_{pp}$, turns ratio = 1:10.

$V_{out} =$ _____

C. $V_{in} = 100\ V_{rms}$. Find V_{out} when the primary and secondary coil have an equal number of turns.

$V_{out} =$ _____

ANSWERS

A. $4\,V_{pp}$ (This is a *step-down transformer.*)

B. $10\,V_{pp}$ (This is a *step-up transformer.*)

C. $100\,V_{rms}$ (This is an *isolation transformer,* which is used to separate or isolate the voltage source from the load electrically.)

9 Almost all electronic equipment operated from 120 volts AC house current requires a transformer to convert the 120 volts AC to a more suitable, lower voltage. Figure 10.7 shows a transformer that steps down 120 volts AC to 28 volts AC.

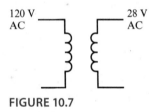

120 V AC — 28 V AC

FIGURE 10.7

QUESTION

Calculate the turns ratio for this transformer. _____

_____

ANSWER

$$TR = \frac{N_P}{N_s} = \frac{120}{28} = 4.3:1$$

10 Figure 10.8 shows an oscilloscope trace of the output waveform from the 28-volt secondary coil.

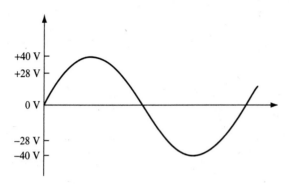

FIGURE 10.8

QUESTIONS

A. Is 28 volts a peak-to-peak or an rms value? _____

_____

B. What is the peak-to-peak value of the 28 volts across the secondary coil? _____

_____

ANSWERS

A. rms

B. 2 × 1.414 × 28 = 79.184 volts

11 Like the 28-volt transformer output value, the 120-volt wall plug value is an rms measurement.

QUESTION

What is the peak-to-peak value of the voltage from the wall plug? _____

_____

_____

ANSWER

Approximately 340 volts

12 The actual voltage measured across the secondary coil of a transformer depends upon where and how you make the measurement. Figure 10.9 illustrates different ways to measure voltage across a 20 V_{pp} secondary coil that has a center tap.

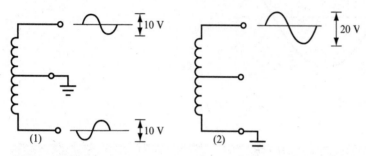

FIGURE 10.9

If the center tap is grounded as shown in diagram (1) of Figure 10.9, then there is 10 V_{pp} AC between each terminal and ground. You can see that the two output waveforms in diagram (1) are out of phase (180 degrees out of phase, in this case) by comparing the two sine waves shown next to the two terminals. If the bottom terminal is grounded as it is in diagram (2) of Figure 10.9 and the center tap is not used, then there is 20 V_{pp} between the top terminal and ground.

QUESTIONS

A. Assume a center-tapped secondary coil is rated at 28 V_{rms} referenced to the center tap. What is the rms voltage output when the center tap is grounded? _____

_____

B. Assume the 28 V_{rms} is the total output voltage across the entire secondary winding. What will be the output voltage between each end of the coil and the center tap? __

_____

C. Assume the output voltage of a center-tapped secondary coil is 15 V_{rms} between each end of the coil and the center tap. What is the peak-to-peak output voltage when the center tap is not connected? _____

_____

ANSWERS

A. 28 V_{rms} between each end of the coil and the center tap.

B. 14 V_{rms} (one half of the total V_{out}).

C. When the center tap is not connected, the output is 30 V_{rms}. Therefore, $V_{pp} = 2 \times 1.414 \times 30 = 84.84$ volts.

13 When the magnetic field induces an AC signal on the secondary coil, there is some loss of power. The percentage of power out of the transformer versus the input power is called the *efficiency* of the transformer. For the sake of this discussion, assume the transformer has an efficiency of 100 percent. Therefore, the output power of the secondary coil equals the power into the primary coil.

Power in = Power out (or $P_{in} = P_{out}$)

However, P = VI. Therefore, the following is true:

$V_{in}I_{in} = V_{out}I_{out}$

You can rearrange this to come up with the following formula:

$$\frac{I_{out}}{I_{in}} = \frac{V_{in}}{V_{out}} = TR$$

QUESTIONS

A. What would be the input current for a transformer if the input power was 12 watts at a voltage of 120 V_{rms}? _____

B. What would be the transformer's output voltage if the turns ratio was 5:1? _____

C. What would be the output current? _____

D. What would be the output power? _____

ANSWERS

In AC power calculations, you must use the rms values of current and voltage.

A. $I_{in} = \dfrac{P_{in}}{V_{in}} = \dfrac{12}{120} = 0.1\,A_{rms}$

B. $V_{out} = \dfrac{V_{in}}{TR} = \dfrac{120}{5} = 24\,V_{rms}$

C. $I_{out} = I_{in}\,(TR) = 0.1 \times 5 = 0.5\,A_{rms}$

D. $P_{out} = V_{out}I_{out} = 24 \times 0.5 = 12$ watts (same as the power in)

INSIDE THE TRANSFORMER

In addition to the turns ratio discussed in this chapter, the design of transformers incorporates a few more aspects. The frequency at which a transformer is expected to operate has a big impact on the design and composition of the core. Transformers with iron cores work well at low frequencies, such as 50 or 60 Hz household AC, and even at audio frequencies. Transformers used at these frequencies are often made of laminated sheets of iron, instead of one solid piece of iron, which increases the electrical resistance of the core, which reduces the eddy current. The *eddy current* is an electrical current induced in the core by the fluctuating magnetic field that reduces the efficiency of transformers.

Reducing the eddy current is especially important at high frequencies. A transformer designed to work in the MHz range requires a core with higher electrical resistance to reduce the eddy current. Therefore, high-frequency transformers have cores made of different materials, such as iron oxides (called *ferrites*) or powdered iron. Transformers are rated for a particular frequency range, which you can find either in the supplier's catalog or in the manufacturer's data sheet.

The maximum power that can pass through a transformer is stated as a *VA rating*. VA stands for "volts x amperes" and is dependent upon factors such as the gauge of wire used. The VA rating makes it easy to calculate the maximum amperage when you know the voltage your circuit requires.

Transformers may also be rated by their input and output impedance at a particular frequency stated in the data sheet.

TRANSFORMERS IN COMMUNICATIONS CIRCUITS

14 In communications circuits, an input signal is often received via a long interconnecting wire (usually called a *line*) that normally has an impedance of 600 ohms. A typical example is a telephone line between two cities.

QUESTION

Communications equipment works best when connected to a load that has the same impedance as the output of the equipment. What output impedance should communications equipment have? _____

_____

ANSWER

600 ohms output impedance, to be connected to a 600-ohm line

15 Because most electronic equipment does not have a 600-ohm output impedance, a transformer is often used to connect such equipment to a line. Often, the transformer is built into the equipment for convenience. The transformer is used to "match" the equipment to the line, as shown in Figure 10.10.

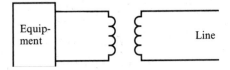

FIGURE 10.10

To work correctly, the output of the transformer secondary coil should have a 600-ohm impedance to match the line. The output impedance of the transformer (measured at the secondary winding) is governed by two things. One of these is the output impedance of the equipment.

QUESTION

What would you expect the other governing factor to be? _____

_____

ANSWER

The turns ratio of the transformer. (The DC resistance of each coil has no effect, and you can ignore it.)

16 Figure 10.11 shows a signal generator with an output impedance of Z_G connected to the primary coil of a transformer. A load impedance of Z_L is connected to the secondary coil.

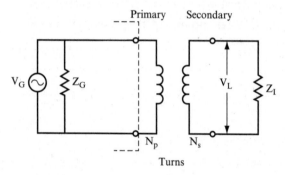

FIGURE 10.11

You know that $P_{in} = P_{out}$ and that $P = V^2/Z$. Therefore, you can write an equation equating the power of the generator to the power of the load in terms of V and Z, as shown here:

$$\frac{V_G^2}{Z_G} = \frac{V_L^2}{Z_L}$$

You can rearrange this equation to give the ratio of the voltages, as shown here:

$$\frac{Z_G}{Z_L} = \left(\frac{V_G}{V_L}\right)^2$$

And, because $V_G = V_{in}$, and $V_L = V_{out}$, and $V_{in}/V_{out} = N_p/N_s$, the following is true:

$$\frac{Z_G}{Z_L} = \left(\frac{V_{in}}{V_{out}}\right)^2 = \left(\frac{N_p}{N_s}\right)^2 = (TR)^2$$

Therefore, the ratio of the input impedance to the output impedance of a transformer is equal to the square of the turns ratio. As you can see in the following question A, you can determine the turns ratio for a transformer that matches impedances between a generator and a load. In this way, the generator "sees" an impedance equal to its own impedance, and the load also "sees" an impedance equal to its own impedance.

For the following problem, a generator has an output impedance of 10 kΩ and produces a 10 V$_{pp}$ (3.53 V$_{rms}$) signal. It will be connected to a 600-ohm line.

QUESTIONS

A. To properly match the generator to the line, what turns ratio is required? _____

_____

B. Find the output voltage across the load. _____

_____

C. Find the load current and power. _____

_____

ANSWERS

A. $TR = \sqrt{\dfrac{Z_G}{Z_L}} = \sqrt{\dfrac{10,000\,\Omega}{600\,\Omega}} = \dfrac{4.08}{1}$ or 4.08 : 1

B. $V_L = \dfrac{V_G}{TR} = \dfrac{10}{4.08} = 2.45\,V_{pp}$, which is 0.866V$_{rms}$

C. $P_{in} = \dfrac{V_G{}^2}{Z_G} = \dfrac{(3.53)^2}{10,000} = 1.25\,mW$

NOTE For the power calculation, you must use the rms value of the voltage.

$I_{in} = \dfrac{P_{in}}{V_{in}} = \dfrac{1.25\,mW}{3.53\,V_{rms}} = 0.354\,mA_{rms}$, which is 1 mA$_{pp}$

$I_L = I_{in}(TR) = 0.354 \times 4.08 = 1.445\,mA_{rms}$, which is 4.08 mA$_{pp}$

$P_L = \dfrac{V_L{}^2}{Z_L} = \dfrac{(0.866)^2}{600} = 1.25\,mW$, which is the same as the input power. This circuit is shown in Figure 10.12.

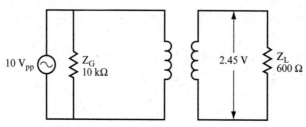

FIGURE 10.12

NOTE   The generator now sees 10 kΩ when it looks toward the load, rather than the actual 600-ohm load. By the same token, the load now sees 600 ohms when it looks toward the source. This condition allows the optimum transfer of power to take place between the source and the load. In practice, however, the optimum condition as calculated here rarely exists. Because it may be impossible to obtain a transformer with a turns ratio of 4.08:1, you would have to select the closest available value, which might be a turns ratio of 4:1. The difference in the turns ratio affects the conditions at the load side, but only slightly.

17 In this problem, you use a transformer to match a generator to a load.

QUESTIONS

A. What turns ratio is required to match a generator that has a 2 kΩ output to a 600-ohm line? _____

_____

B. If the generator produces 1 V_{pp}, what is the voltage across the load? _____

_____

ANSWERS

A. TR = 1.83

B. $V_L = 0.55\ V_{pp}$

18 In this problem you use a transformer to match a generator to a 2 kΩ load.

QUESTIONS

A. What turns ratio is required to match a 2 kΩ load with a source that has an output impedance of 5 kΩ? _____

_____

B. If the load requires a power of 20 mW, what should the source be? (First, find the voltage across the load.) _____

_____

C. What are the primary and secondary currents and the power supplied by the source to the primary side of the transformer? _____

_____

ANSWERS

A. $TR = 1.58$

B. $V_L = \sqrt{P_L \times Z_L} = \sqrt{20\,mW \times 2\,k\Omega} = 6.32\,V_{rms}$; and

$V_G = V_L \times TR = 6.32\,V_{rms} \times 1.58 = 10\,V_{rms}$

C. $I_L = 3.16\,mA_{rms}$, $I_p = 2\,mA_{rms}$, $P_{in} = 20\,mW$

SUMMARY AND APPLICATIONS

In this chapter, you learned about the following topics related to transformers:

- The principles that allow an AC signal to be induced in a secondary coil
- How the AC voltage across the secondary coil can be stepped up or down depending upon the turns ratio of the transformer
- The use of a center tap to produce various voltages from a transformer

- The use of transformers to match impedances between a generator and a load
- That transformers can cause the output signal to be inverted (out of phase) from the input signal

SELF-TEST

These questions test your understanding of the material in this chapter. Use a separate sheet of paper for your diagrams or calculations. Compare your answers with the answers provided following the test.

1. How is a transformer constructed? _____

_____

2. What type of signal is used as an input to a transformer? _____

_____

3. If a sine wave is fed into a transformer shown in Figure 10.13, what does the output waveform look like? _____

_____

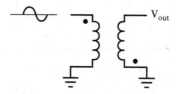

FIGURE 10.13

4. What is meant by the term *turns ratio*? _____

_____

5. If $V_{in} = 1\ V_{pp}$ and TR = 2, what is the output voltage?

V_{out} _____

6. $V_{in} = 10\ V_{pp}$ and $V_{out} = 7\ V_{pp}$, what is the turns ratio?

TR = _____

_____

7. In the center-tapped secondary winding shown in Figure 10.14, the voltage between points A and B may be expressed as $V_{A-B} = 28\ V_{pp}$. What is the voltage between C and A? _____

_____

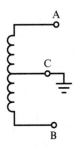

FIGURE 10.14

8. In the center-tapped secondary winding shown in Figure 10.14, the voltage between points B and C is $V_{B-C} = 5\ V_{rms}$. What is the peak-to-peak voltage between A and B? _____

_____

9. If $I_{in} = 0.5\ A_{rms}$ and $I_{out} = 2.0\ A_{rms}$, what is the turns ratio? _____

_____

10. Is the transformer in problem 9 a step-up or a step-down transformer? _____

_____

11. If $Z_L = 600$ ohms and $Z_G = 6\ k\Omega$, find the turns ratio.

TR = _____

_____

12. If $Z_L = 1\ k\Omega$ and the turns ratio is 10:1, what is the generator impedance?

$Z_G =$ _____

_____

ANSWERS TO SELF-TEST

If your answers do not agree with those given here, review the problems indicated in parentheses before you go to Chapter 11.

1.	Two coils of wire wound around a magnetic core (such as iron or ferrite).	(problem 1)
2.	An AC voltage—DC does not work.	(problem 2)
3.	An inverted sine wave.	(problem 3)
4.	The ratio of the turns in the primary winding to the number of turns in the secondary coil.	(problem 6)
5.	$V_{out} = 0.5$ volts.	(problem 7)
6.	TR = 1.43:1.	(problem 7)
7.	$V_{C-A} = 14\ V_{pp.}$	(problem 12)
8.	$V_{A-B} = 14.14\ V_{pp.}$	(problem 12)
9.	TR = 4:1.	(problem 13)
10.	It is a step-down transformer. The voltage is lower (stepped down) in the secondary coil than in the primary coil if the current in the secondary coil is higher than the current in the primary coil. This maintains the same power on either side of the transformer.	(problem 13)
11.	TR = 3.2:1.	(problem 16)
12.	$Z_G = 100\ k\Omega.$	(problem 16)

Power Supply Circuits

A power supply is incorporated into many electronic devices. Power supplies convert the 120 volts AC from a wall plug to a DC voltage, providing power for all types of electronic circuits.

Power supply circuits are simple in principle, and those shown in this chapter have been around for many years. Because power supplies incorporate many of the features covered in this book, they make an excellent conclusion to your study of basic electronics.

Diodes are a major component in power supplies. Learning how AC signals are affected by diodes is fundamental to your understanding of how power supplies work. Therefore, this chapter begins with a brief discussion of diodes in AC circuits.

Throughout this chapter, diagrams show how AC signals are affected by diodes and other components in power supply circuits. If you have an oscilloscope, you can bread-board the circuits and observe these waveforms.

When you complete this chapter, you will be able to do the following:

- Describe the function of diodes in AC circuits.
- Identify at least two ways to rectify an AC signal.
- Draw the output waveforms from rectifier and smoothing circuits.
- Calculate the output voltage from a power supply circuit.
- Determine the appropriate component values for a power supply circuit.

DIODES IN AC CIRCUITS PRODUCE PULSATING DC

1 You can use diodes for several purposes in AC circuits, where their characteristic of conducting in only one direction is useful.

QUESTIONS

Assume that you apply + 20 volts DC at point A of the circuit, as shown in Figure 11.1.

A. What is the output voltage at point B? _____

_____

B. Suppose that you now apply + 10 volts DC at point B. What is the voltage at point A? _____

_____

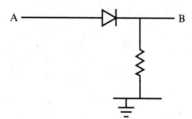

FIGURE 11.1

ANSWERS

A. 20 volts DC (Ignore, for now, the voltage drop of 0.7 volt across the diode.)

B. 0 volts (The diode is reverse-biased.)

2 Figure 11.2 shows the circuit in Figure 11.1 with a 20 V_{pp} AC input signal centered at $+20$ volts DC.

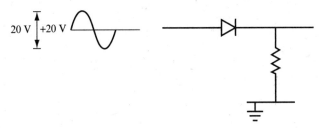

FIGURE 11.2

QUESTIONS

A. What are the positive and negative peak voltages of the input signal? _____

_____

B. What is the output waveform of this circuit? _____

_____

ANSWERS

A. Positive peak voltage is 20 volts + 10 volts = 30 volts. Negative peak voltage is 20 volts − 10 volts = 10 volts.

B. The diode is always forward-biased, so it always conducts. Thus, the output wave-form is exactly the same as the input waveform.

3 Figure 11.3 shows a circuit with 20 V_{pp} AC input signal centered at 0 volts DC.

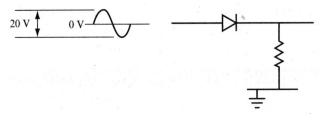

FIGURE 11.3

QUESTIONS

A. What are the positive and negative peak voltages of the input signal? _____

_____

B. For the positive half wave of the input, draw the output waveform on the blank graph provided in Figure 11.4.

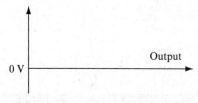

FIGURE 11.4

ANSWERS

A. Positive peak voltage is +10 volts. Negative peak voltage is −10 volts.

B. See Figure 11.5.

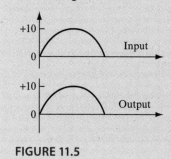

FIGURE 11.5

4 When the input is negative, the diode in the circuit shown in Figure 11.3 is reverse-biased. Therefore, the output voltage remains at 0 volts.

QUESTION

Figure 11.6 shows the input waveform for the circuit shown in Figure 11.3. Draw the output waveform on the blank graph provided in Figure 11.6.

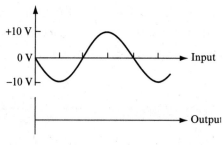

FIGURE 11.6

ANSWER

See Figure 11.7.

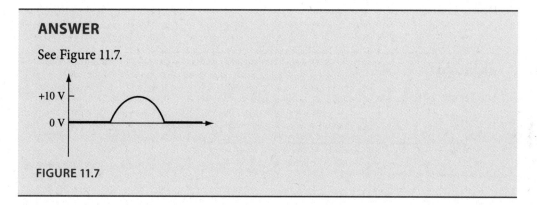

FIGURE 11.7

5 Figure 11.7 shows the output waveform of the circuit shown in Figure 11.3, for one complete cycle of the input waveform.

QUESTION

Now, draw the output waveform for three complete cycles of the input waveform shown in Figure 11.6. Use a separate sheet of paper for your drawing.

ANSWER

See Figure 11.8.

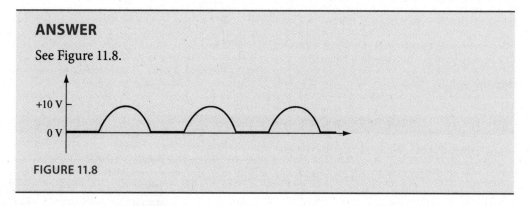

FIGURE 11.8

6 When the diode is connected in the opposite direction, it is forward-biased and, therefore, conducts current when the input signal is negative. In this case, the diode is reverse-biased when the input signal is positive. Therefore, the output waveform is inverted from the output waveform shown in Figure 11.8.

QUESTION

On a separate sheet of paper, draw the output waveform for three input cycles, assuming that the diode is connected in the opposite direction from the diode shown in Figure 11.3.

ANSWER

See Figure 11.9.

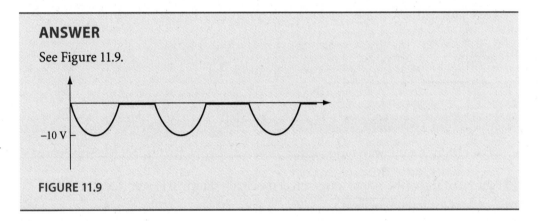

FIGURE 11.9

7 Figure 11.10 shows a circuit with a 20 V_{pp} AC input signal centered at –20 volts DC.

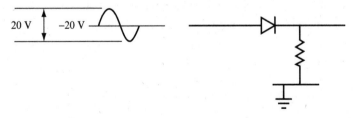

FIGURE 11.10

QUESTIONS

A. When is the diode forward-biased? _____

_____

B. What is the output voltage? _____

_____

ANSWERS

A. Never because the voltage that results from adding the AC and DC signals ranges from −10 volts to −30 volts. Therefore, the diode is always reverse-biased.

B. A constant 0 volts.

8 As you have seen, a diode passes either the positive or negative portion of an AC voltage waveform, depending on how you connect it in a circuit. Therefore, the AC input signal is converted to a pulsed DC output signal, a process called *rectification*. A circuit that converts either the positive or negative portion of an AC voltage waveform to a pulsed DC output signal is called a *half-wave rectifier*.

QUESTION

Refer to the output waveforms shown in Figures 11.8 and 11.9. Do these waveforms represent positive DC voltage pulses or negative DC voltage pulses? _____

_____

ANSWER

The waveform in Figure 11.8 represents positive pulses of DC voltage. The waveform in Figure 11.9 represents negative pulses of DC voltage.

9 The circuit shown in Figure 11.11 shows a diode connected to the secondary coil of a transformer.

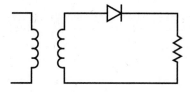

FIGURE 11.11

QUESTIONS

A. How does the diode affect the AC signal? _____

_____

B. Draw the waveform of the voltage across the load for the circuit shown in Figure 11.11 if the secondary coil of the transformer has a 30 V_{pp} AC output signal centered at 0 volts DC. Use a separate sheet of paper for your drawing.

ANSWERS

A. The AC signal is rectified.

B. See Figure 11.12. This type of circuit (called a *half-wave rectifier*) produces an output waveform containing either the positive or negative portion of the input waveform.

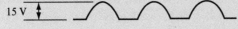

15 V

FIGURE 11.12

10 Figure 11.13 shows the waveforms at each end of a center-tap transformer secondary coil. Diode D_1 rectifies the waveform shown at point A, and diode D_2 rectifies the waveform shown at point B.

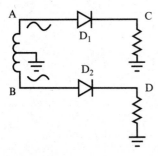

FIGURE 11.13

QUESTIONS

A. Which diode conducts during the first half of the cycle? _____

_____

B. Which diode conducts during the second half of the cycle? _____

_____

C. Draw the input waveforms (points A and B), and underneath draw each output waveform (points C and D). Use a separate sheet of paper for your drawing.

ANSWERS

A. During the first half of the cycle, D_1 is forward-biased and conducts current. D_2 is reverse-biased and does not conduct current.

B. During the second half of the cycle, D_2 is forward-biased and conducts current. D_1 is reversed-biased and does not conduct current.

C. See Figure 11.14.

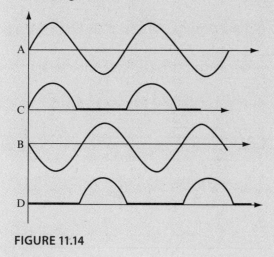

FIGURE 11.14

11 Figure 11.15 shows a circuit in which diodes connected to the ends of a center-tap transformer are connected to ground through a single resistor. The output voltage waveforms from both diodes are therefore applied across one load resistor. This type of circuit is called a *full-wave rectifier.*

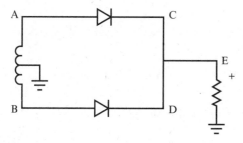

FIGURE 11.15

QUESTION

On a separate sheet of paper, draw the waveform representing the voltage at point E in the circuit, as shown in Figure 11.15. (This waveform is a combination of the waveforms at points C and D shown in Figure 11.14.)

ANSWER

See Figure 11.16.

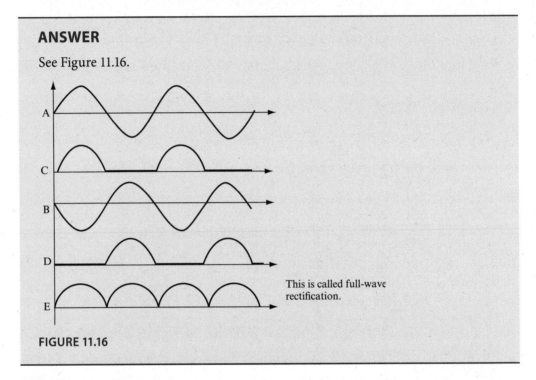

This is called full-wave rectification.

FIGURE 11.16

12 Full-wave rectification of AC allows a much "smoother" conversion of AC to DC than half-wave rectification.

Figure 11.17 shows a full-wave rectifier circuit that uses a transformer with a two-terminal secondary coil, rather than a center-tapped secondary coil. This type of circuit is called a *bridge rectifier*.

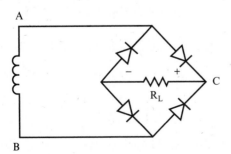

FIGURE 11.17

QUESTION

How does this circuit differ from the circuit shown in Figure 11.15? _____

_____

ANSWER

This circuit has no center tap on the secondary coil, and it uses four diodes.

13 Figure 11.18 shows the direction of current flow when the voltage at point A is positive.

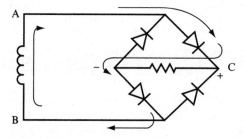

FIGURE 11.18

Figure 11.19 shows the direction of current flow when the voltage at point B is positive.

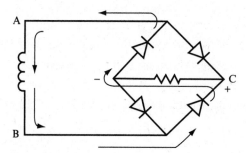

FIGURE 11.19

Notice that the direction of current through the load resistor is the same in both cases.

QUESTIONS

A. Through how many diodes does the current travel in each conduction path? _____

_____

B. Draw the voltage waveform at point C. Use a separate sheet of paper for your drawing.

ANSWERS

A. Two diodes in each case.

B. See Figure 11.20.

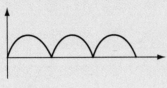

FIGURE 11.20

PROJECT 11.1: The Full-Wave Rectifier

OBJECTIVE

The objective of this project is to compare the outputs of the two types of full-wave rectifiers.

GENERAL INSTRUCTIONS

You set up two circuits. One of the circuits is a full-wave rectifier containing a center-tapped transformer and two diodes. The other circuit is a bridge rectifier containing a transformer and four diodes. After each circuit is set up, you apply a 20 V_{pp} signal to the primary side of the transformer. Then you use your oscilloscope to look at the waveform across the load resistor, and measure the peak voltage of each waveform.

Parts List

You need the following equipment and supplies:

❑ Six 1N4001 diodes.

❑ Two 10 kΩ, 0.25-watt resistors.

❑ One audio transformer with impedance of both the primary and secondary coil rated at 600 Ω (with equal impedance for the primary and secondary coils, the turns ratio will be 1:1), and a center-tapped secondary coil. You can use a transformer with a center tap on both the primary and secondary coil; just don't connect a center tap that's not called for in the schematic.

❑ Two breadboards.

❑ One function generator.

❑ One oscilloscope.

STEP-BY-STEP INSTRUCTIONS

Set up Circuit # 1, the full-wave rectifier circuit shown in Figure 11.21. If you have some experience in building circuits, this schematic (along with the previous parts list) should provide all the information you need to build the circuit. If you need a bit more help building the circuit, look at the photos of the completed circuit in the "Expected Results" section.

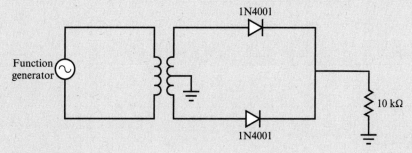

FIGURE 11.21

Carefully check your circuit against the diagram.

After you check your circuit, follow these steps and record your measurements in the blank table following the steps.

1. Connect the oscilloscope probe for Channel 1 to a jumper wire connected to the end of the resistor nearest the diodes. Connect the ground clip to a jumper wire attached to the ground bus.

2. Connect the oscilloscope probe for Channel 2 to a jumper wire that is connected to one end of the primary coil. This should be the same end to which you've connected the red lead from the function generator. Connect the ground clip for Channel 2 to a jumper wire that is connected to the other end of the primary coil. This should be the end to which you've connected the black lead from the function generator.

3. Set the function generator to generate a 1-kHz sine wave with 20 V_{pp}.

4. Measure and record V_p for the signal across the resistor.

5. Set up Circuit # 2; the bridge rectifier circuit shown in Figure 11.22. Use the same transformer you used in Circuit #1. You do not connect the center tap on the secondary coil in this circuit.

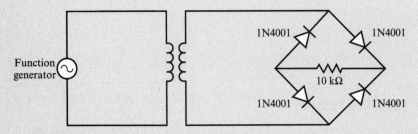

FIGURE 11.22

6. Connect the oscilloscope probe for Channel 1 to a jumper wire connected to one end of the resistor. Connect the ground clip for Channel 1 to a jumper wire connected to the other end of the resistor.

7. Repeat steps 2 through 4.

Circuit #	V_{PP} (Primary Coil)	V_P (Load Resistor)
1		
2		

EXPECTED RESULTS

Figure 11.23 shows the breadboarded full-wave rectifier (Circuit # 1).

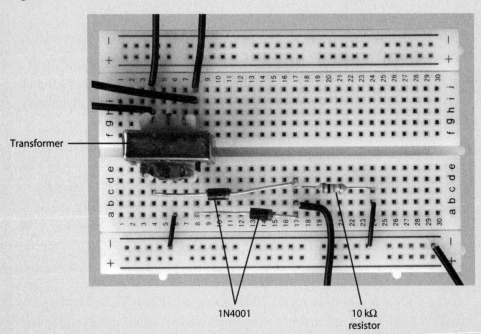

FIGURE 11.23

Figure 11.24 shows a function generator and oscilloscope attached to Circuit #1.

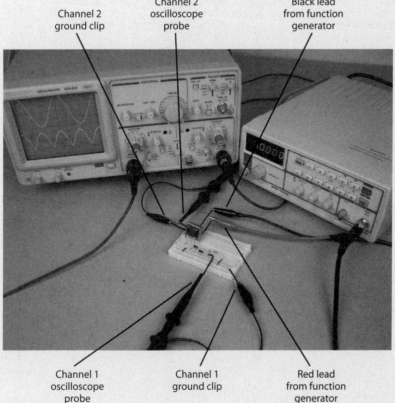

FIGURE 11.24

In Figure 11.25, the signal across the primary coil is represented by the upper wave-form, and the signal across the resistor is represented by the lower waveform. Read the number of divisions for the peak-to-peak voltage of the upper waveform, and multiply it by the corresponding VOLTS/DIV setting to determine V_{pp} for the signal across the primary coil. Read the number of divisions for the peak voltage of the lower waveform, and multiply it by the corresponding VOLTS/DIV setting to determine V_p for the signal across the resistor.

As you measure V_{pp} and V_p, you may need to adjust the TIME/DIV, VOLTS/DIV, and vertical POSITION controls on the oscilloscope. The controls shown in Figure 11.26 are adjusted to measure V_{pp} in the primary coil and V_p across the load resistor for Circuit # 1.

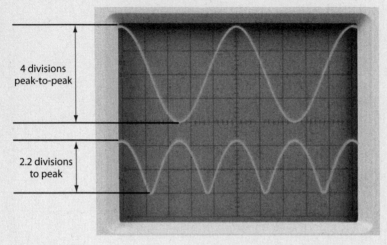

4 divisions
peak-to-peak

2.2 divisions
to peak

FIGURE 11.25

Time/div control set
to 0.2 msec/div

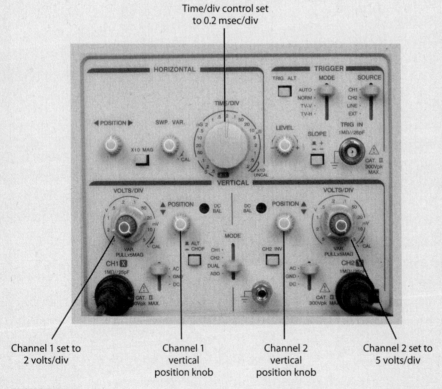

Channel 1 set to
2 volts/div

Channel 1
vertical
position knob

Channel 2
vertical
position knob

Channel 2 set to
5 volts/div

FIGURE 11.26

Figure 11.27 shows the breadboarded bridge rectifier (Circuit # 2).

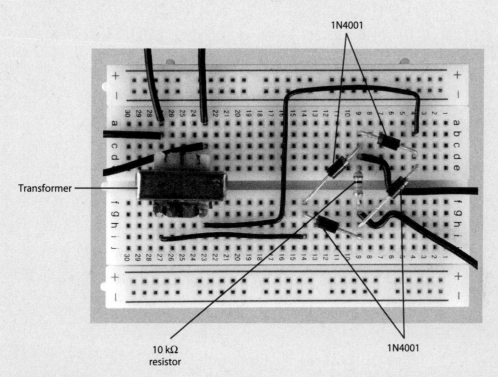

1N4001

Transformer

10 kΩ
resistor

1N4001

FIGURE 11.27

Figure 11.28 shows a function generator and oscilloscope attached to Circuit #2.

In Figure 11.29, the signal across the primary coil is represented by the upper wave-form, and the signal across the resistor is represented by the lower waveform. Read the number of divisions for the peak-to-peak voltage of the upper waveform, and multiply it by the corresponding VOLTS/DIV setting to determine V_{pp} for the signal across the primary coil. Read the number of divisions for the peak voltage of the lower waveform, and multiply it by the corresponding VOLTS/DIV setting to determine V_p for the signal across the resistor.

As you measure V_{pp} and V_p, you may need to adjust the TIME/DIV, the VOLTS/DIV, and the vertical POSITION controls on the oscilloscope. The controls shown in Figure 11.30 are adjusted to measure V_{pp} in the primary coil and V_p across the load resistor for Circuit # 2.

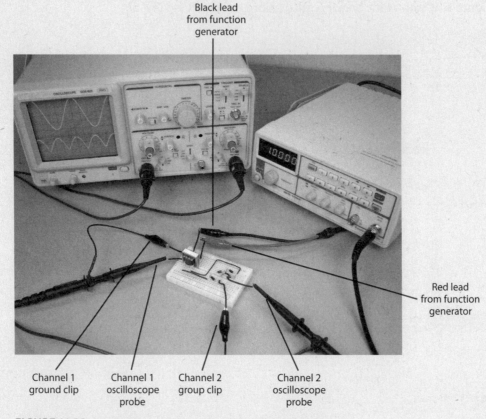

Black lead
from function
generator

Red lead
from function
generator

Channel 1
ground clip

Channel 1
oscilloscope
probe

Channel 2
group clip

Channel 2
oscilloscope
probe

FIGURE 11.28

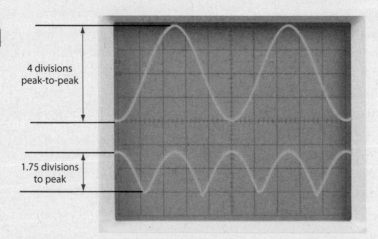

4 divisions
peak-to-peak

1.75 divisions
to peak

FIGURE 11.29

Time/div control set
to 0.2 msec/div

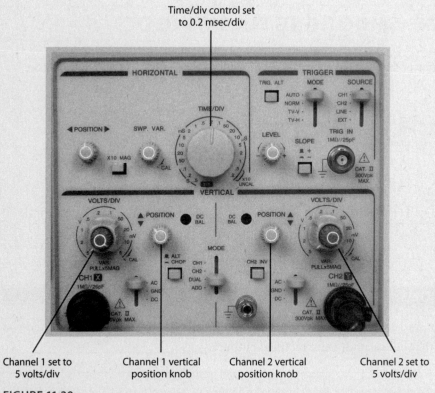

Channel 1 set to 5 volts/div   Channel 1 vertical position knob   Channel 2 vertical position knob   Channel 2 set to 5 volts/div

FIGURE 11.30

Your values should be close to those shown in the following table.

Circuit #	V_{pp} (Primary Coil)	V_p (Load Resistor)
1	20 volts	4.4 volts
2	20 volts	8.8 volts

V_p across the load resistor for Circuit # 2 is twice the value of V_p for Circuit #1. That's because Circuit #1 uses a center-tapped secondary coil that supplies a 10 V_{pp} sine wave to the diodes, whereas the secondary coil in Circuit #2 supplies a 20 V_{pp} sine wave. In each circuit, the 0.6 to 0.7 voltage drop that occurs as the signal passes through each diode causes V_p across the load resistor to be slightly lower than half the V_{pp} supplied by the secondary coil.

What this chapter has explored to this point is how AC is turned into pulsating DC. In fact, rectified AC is often called *pulsating DC*. The next step in your understanding of power supplies is to learn how you turn pulsating DC into level DC.

LEVEL DC (SMOOTHING PULSATING DC)

14 A basic power supply circuit can be divided into four sections, as shown in Figure 11.31.

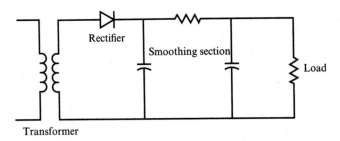

FIGURE 11.31

QUESTIONS

A. If you use a center-tap transformer in a power supply, how many diodes would you need to produce a full-wave rectified output? _____

_____

B. Will the power supply circuit shown in Figure 11.31 result in full- or half-wave rectification? _____

_____

C. What type of output will the rectifier section of the power supply circuit shown in Figure 11.31 produce? _____

_____

ANSWERS

A. Two

B. Half-wave

C. Pulsating DC

15 The function of the smoothing section of a power supply circuit is to take the pulsating DC (PDC) and convert it to a "pure" DC with as little AC "ripple" as possible. The smoothed DC voltage, shown in the illustration on the right in Figure 11.32, is then applied to the load.

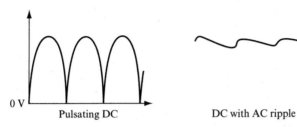

FIGURE 11.32

The load (which is "driven" by the power supply) can be a simple lamp or a complex electronic circuit. Whatever load you use, it requires a certain voltage across its terminals and draws a current. Therefore, the load has a resistance.

Usually, the voltage and current required by the load (and, hence, its resistance) are known, and you must design the power supply to provide that voltage and current.

To simplify the circuit diagrams, you can treat the load as a simple resistor.

QUESTIONS

A. What does the smoothing section of a power supply do? _____

_____

B. What is connected to a power supply, and what can you treat it like? _____

_____

ANSWERS

A. The smoothing section converts the pulsating DC to a "pure" DC.

B. A load such as a lamp or an electronic circuit is connected to a power supply. In most cases, you can treat the load as you do a resistor.

16 Figure 11.33 shows a power supply circuit with a resistor as the load.

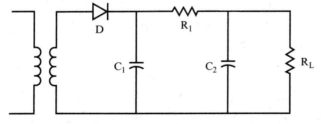

FIGURE 11.33

QUESTIONS

Look at the circuit shown in Figure 11.33 and answer the following questions:

A. What type of secondary coil is used? _____

_____

B. What type of rectifier is used? _____

_____

C. What components make up the smoothing section? _____

_____

D. What output would you expect from the rectifier section? _____

_____

ANSWERS

A. A secondary coil with no center tap

B. A single-diode, half-wave rectifier

C. A resistor and two capacitors (R_1, C_1, and C_2)

D. Half-wave pulsating DC

17 Figure 11.34 shows the output waveform from the rectifier portion of the power supply circuit shown in Figure 11.33.

FIGURE 11.34

This waveform is the input to the smoothing section of the power supply circuit. Use one of the DC pulses (shown in Figure 11.35) to analyze the effect of the smoothing section on the waveform.

FIGURE 11.35

As the voltage level of the DC pulse rises to its peak, the capacitor C_1 is charged to the peak voltage of the DC pulse.

When the input DC pulse drops from its peak voltage back to 0 volts, the electrons stored on capacitor C_1 discharge through the circuit. This maintains the voltage across the load resistor at close to its peak value, as shown in Figure 11.36. The DC pulse to the right of the diode stays at the peak voltage, even though V_{in} drops to zero.

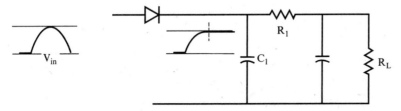

FIGURE 11.36

QUESTION

What discharge path is available for the capacitor C_1? _____

_____

ANSWER

The diode is not conducting, so the capacitor cannot discharge through the diode. The only possible discharge path is through R_1 and the load R_L.

18 If no further pulses pass through the diode, the voltage level drops as the capacitor discharges, resulting in the waveform shown in Figure 11.37.

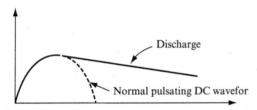
Discharge

Normal pulsating DC wavefor

FIGURE 11.37

However, if another pulse passes through the diode before the capacitor is discharged, the resulting waveform looks like that shown in Figure 11.38.

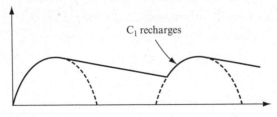

FIGURE 11.38

The capacitor discharges only briefly before the second pulse recharges it to peak value. Therefore, the voltage applied to the load resistor drops only a small amount.

Applying further pulses can produce this same recharging effect again and again. Figure 11.39 shows the resulting waveform.

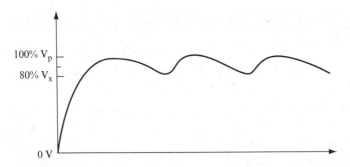

FIGURE 11.39

The waveform in Figure 11.39 has a DC level with an AC ripple, which varies between V_p and V_x. If you choose values of C_1, R_1, and R_L that produce a discharge time constant for C_1 equal to about 10 times the duration of an input pulse, V_x will be approximately 80 percent of V_p.

If the discharge time you select is greater than 10 times the duration of an input pulse, the smoothing effect minimizes the AC ripple. A time constant of 10 times the pulse duration results in practical design values that are used throughout this chapter.

> **NOTE** The smoothing section of a power supply circuit is sometimes referred to as a *low-pass filter*. Though such a circuit can function as a low-pass filter, in the case of a power supply circuit converting AC to DC, it is the release of electrons

by the capacitor that is primarily responsible for leveling out the pulsating DC. For that reason, this discussion uses the term *smoothing section*.

QUESTION

Estimate the average DC output level of the waveform shown in Figure 11.39. _____

ANSWER

Approximately 90 percent of V_p

19 The output from the secondary coil of the circuit shown in Figure 11.40 is a 28 V_{rms}, 60 Hz sine wave. For this circuit, you need 10 volts DC across the 100-ohm load resistor.

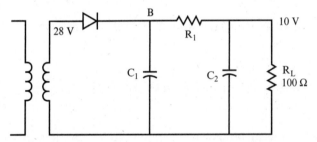

FIGURE 11.40

QUESTION

What is the peak voltage out of the rectifier? _____

ANSWER

The transformer secondary coil delivers 28 V_{rms}, so

$$V_p = \sqrt{2} \times V_{rms} = 1.414 \times 28V_{rms} = 39.59 \text{ volts}$$

or about 40 volts.

20 Figure 11.41 shows the waveform after the diode has rectified the sine wave for the half-way rectifier circuit shown in Figure 11.40.

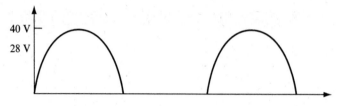

FIGURE 11.41

QUESTION

Calculate the duration of one pulse. _____

_____

ANSWER

60 Hz represents 60 cycles (that is, *wavelengths*) in 1 second. Therefore, one wavelength lasts for 1/60 second.

1/60 second = 1000/60 milliseconds = 16.67 ms

Therefore, the duration of a pulse, which is half a wavelength, is 8.33 ms.

21 The average DC voltage at point B in the circuit shown in Figure 11.40 is approximately 90 percent of the peak value of the sine wave from the secondary coil, or $V_B = 0.9 \times 40$ volts = 36 volts. R_1 and R_L act as a voltage divider to reduce the 36-volt DC level to the required 10 volts DC at the output.

QUESTION

Using the voltage divider formula, calculate the value of R_1 that will result in 10 volts DC across the 100-ohm load resistor. _____

_____

ANSWER

$$V_{out} = \frac{V_{in} R_L}{(R_1 + R_L)}$$

$$10 = \frac{36 \times 100}{(R_1 + 100)}$$

Therefore, R_1 = 260 ohms

22 Figure 11.42 shows the half-wave rectifier circuit with the 260-ohm value you calculated for R_1.

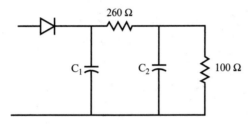

FIGURE 11.42

Now, choose a value for C_1 that produces a discharge time through the two resistors equal to 10 times the input wave duration.

QUESTIONS

A. How long should the discharge time constant be for the circuit in Figure 11.42? Refer to problems 18 and 20. _____

_____

B. Given the time constant, calculate the value of C_1. _____

_____

ANSWERS

A. The time constant should be 10 times the pulse duration (8.33 ms), so

$\tau = 10 \times 8.33$ ms = 83.3 ms or 0.083 seconds

B. $\tau = R \times C = (R_1 + R_L) \times C_1 = 360 \times C_1$

Therefore, 0.0833 = 360 × C_1, or C_1 = 230 μF

23 Figure 11.43 shows voltage waveforms at various points in the half-wave rectifier circuit.

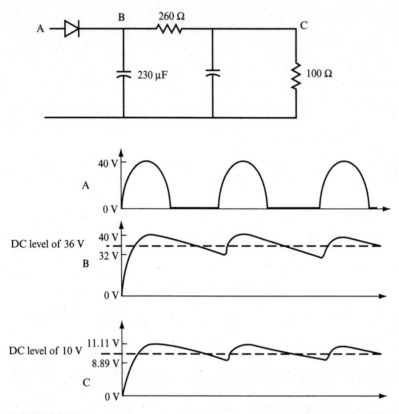

FIGURE 11.43

QUESTIONS

A. What happens to the DC output voltage between points B and C in this circuit? ___

_____

B. What happens to the waveform between points A and C in the circuit? _____

_____

ANSWERS

A. The voltage has been reduced from 36 volts to 10 volts.

B. The waveform has changed from pulsating DC to a 10-volt DC level with an AC ripple.

24 In most cases, the level of the AC ripple is still too high, and further smoothing is required. Figure 11.44 shows the portion of the half-wave rectifier circuit that forms a voltage divider using R_1, and the parallel combination of R_L and C_2. This voltage divider reduces the AC ripple and the DC voltage level.

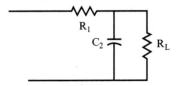

FIGURE 11.44

Choose a value for C_2 that causes the capacitor's reactance (X_{C2}) to be equal to or less than one-tenth of the resistance of the load resistor. C_2, R_1, and R_2 form an AC voltage divider. As discussed in problem 26 of Chapter 6, "Filters," choosing such a value for C_2 simplifies the calculations for an AC voltage divider circuit containing a parallel resistor and capacitor.

QUESTIONS

A. What should the value of X_{C2} be? _____

_____

B. What is the formula for the reactance of a capacitor? _____

_____

C. What is the frequency of the AC ripple? _____

_____

D. Calculate the value of the capacitor C_2. _____

_____

ANSWERS

A. $X_{C2} = R_L/10 = 100/10 = 10\ \Omega$ or less

B. $X_C = \dfrac{1}{2\pi fC}$

C. 60 Hz. This is identical to the frequency of the sine wave output from the transformer's secondary coil.

D. Solving the reactance formula for C results in the following:

$$C_2 = \frac{1}{2\pi fX_c} = \frac{1}{2 \times \pi \times 60\,\text{Hz} \times 10\,\Omega} = 265\,\mu\text{F}$$

25 Figure 11.45 shows the half-wave rectifier circuit with all capacitor and resistor values.

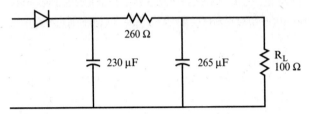

FIGURE 11.45

Because X_{C_2} is one-tenth of R_L, you can ignore R_L in AC voltage divider calculations. Figure 11.46 shows the resulting AC voltage divider circuit.

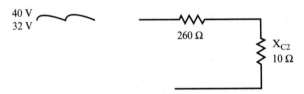

FIGURE 11.46

QUESTIONS

A. What is the peak-to-peak voltage at the input to the AC voltage divider? _____

_____

B. Find the AC ripple output across R_L using the AC voltage divider formula discussed in problem 26 of Chapter 6. _____

_____

ANSWERS

A. $V_{PP} = V_p - V_x = 40 \text{ volts} - 32 \text{ volts} = 8\ V_{PP}$

B. $AC\,V_{out} = (AC\,V_{in}) \times \dfrac{X_{C2}}{\sqrt{X_{C2}{}^2 + R_1{}^2}}$

$AC\,V_{out} = 8 \times \dfrac{10}{\sqrt{(10^2 + 260^2)}} = 0.31\,V_{PP}$

NOTE   This result means that the addition of C_2 lowers the AC ripple shown by curve C in Figure 11.43, with peak values of 11.11 and 8.89, to values of 10.155 and 9.845 volts. This represents a lower ripple at the output. Hence, C_2 aids the smoothing of the 10 volts DC at the output.

26 You can apply the calculations you performed for a half-wave rectifier circuit in the last few problems to a full-wave rectifier circuit. In the next few problems, you calculate the values of R_1, C_1, and C_2 required to provide 10 volts DC across a 100-ohm load for a full-wave rectifier circuit with a 28 V_{rms} sine wave supplied by the secondary coil of a transformer.

Figure 11.47 shows the output waveform from the rectifier section of the circuit.

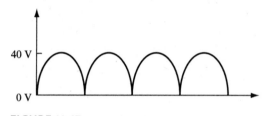

FIGURE 11.47

Figure 11.48 shows the waveform that results from using a smoothing capacitor.

FIGURE 11.48

If the discharge time constant C_1 is 10 times the period of the waveform, V_x is approximately 90 percent of V_p. The average DC level is approximately 95 percent of V_p.

QUESTIONS

A. What is the average DC level for the half-wave rectifier at point B in Figure 11.43? _

_____

_____

B. What is the average DC level for the waveform in Figure 11.48 given that $V_p = 40$ volts? _____

_____

C. Why does a full-wave rectifier have a higher average DC level than a half-wave rectifier? _____

_____

ANSWERS

A. 36 volts, which is 90 percent of V_p.

B. 38 volts, which is 95 percent of V_p.

C. The slightly higher values occur because the capacitor does not discharge as far with full-wave rectification, and, as a result, there is slightly less AC ripple. Therefore, V_x is higher and the average DC level is higher.

27 You can use the method for calculating the value of R_1 for a half-wave rectifier (see problem 21) to calculate the value of R_1 for a full-wave rectifier.

QUESTION

Calculate the value of R_1 when $R_L = 100\ \Omega$, $V_{in} = 38$ volts, and the required voltage across R_L is 10 volts. _____

_____

ANSWER

$$V_{out} = 10\,volts = \frac{V_{in}R_L}{(R_1 + R_L)} = \frac{38 \times 100}{(R_1 + 100)}$$

Therefore, $R_1 = 280$ ohms.

28 You can also use the method for calculating the value of C_1 for a half-wave rectifier (see problem 22) to calculate the value of C_1 for a full-wave rectifier.

QUESTION

Calculate the value of C_1. _____

_____

ANSWER

With a time constant of $\tau = 83.3$ ms. and a discharge resistance of $R_1 + R_L = 380$ ohms, $C_1 = 220\ \mu F$.

29 You can use the voltage divider equation to find the amount of AC ripple across the load resistor for a full-wave rectifier with $R_1 = 280\ \Omega$ and $R_L = 100\ \Omega$. For $V_p = 40$ volts, the calculation results in 10.52 volts. For $V_x = 36$ volts, the calculation results in 9.47 volts. Therefore, the voltage levels at the load resistor vary between 10.52 volts and 9.47 volts, with an average DC level of 10 volts. You can reduce the AC ripple by adding a second capacitor in parallel with the load resistor.

QUESTIONS

Use the method for calculating the value of C_2 for the half-wave rectifier in problem 24.

A. Calculate the reactance of the second capacitor (C_2). _____

_____

B. Calculate the value of C_2. (The frequency of the AC ripple for the full-wave rectifier is 120 Hz.) _____

_____

ANSWERS

A. The reactance should be one-tenth (or less) of the load resistance. Therefore, it should be 10 ohms or less.

B. $C_2 = \dfrac{1}{2\pi f X_C} = \dfrac{1}{2 \times \pi \times 120\,\text{Hz} \times 10\,\Omega} = 135\,\mu F$

30 The AC ripple at the first smoothing capacitor ranges from 36 volts to 40 volts. The AC ripple at the load ranges from 9.47 volts to 10.52 volts when there is only one capacitor in the circuit.

QUESTION

Calculate the upper and lower values of the AC ripple at the output if you use a second capacitor with a value of 135 μF in parallel with the load resistor. You can use the same formulas as those for the half-wave rectifier in problem 25. X_{C_2} is 10 Ω from problem 29; R_1 = 280 from problem 27; and AC V_{in} = V_p − V_x = (40 volts − 36 volts) = 4 V_{pp}. _____

ANSWER

$$AC\,V_{out} = (AC\,V_{in}) \times \frac{X_{C2}}{\sqrt{(R_1^2 + X_{C2}^2)}} = 0.143\,V_{pp}$$

The result of approximately 0.14 V_{pp} means that the output will now vary from 10.07 to 9.93 volts. This shows that the second capacitor lowers the ripple significantly. The AC ripple is less than half of the ripple shown for the half-wave rectifier in problem 25. In other words, a full-wave rectifier produces a smoother DC output than a half-wave rectifier.

31 Figure 11.49 shows a full-wave rectifier circuit with an output voltage of 5 volts across a 50 Ω load resistor. Use the following steps to calculate the values of the other components.

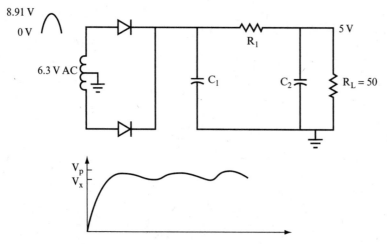

FIGURE 11.49

QUESTIONS

A. What are V_p, V_x, and the DC level at the first capacitor? _____

B. Calculate the value of R_1 required to make the DC level at the output 5 volts. _____

_____

C. Calculate the value of C_1. _____

_____

D. Calculate the value of C_2. _____

_____

E. What is the amount of AC ripple at the output? _____

_____

F. Draw the final circuit showing the calculated values. Use a separate sheet of paper for your drawing.

ANSWERS

A. $V_p = 6.3 \times \sqrt{2} = 8.91$ volts, $V_x = 90$ percent of $V_p = 8.02$ volts.

The DC level is 95 percent of V_p, which is 8.46 volts.

B. About 35 ohms.

C. 980 µF.

D. Using $X_{C2} = 5$ ohms and 120 Hz, $C_2 = 265$ µF.

E. At the input to the smoothing section, the AC variation is 8.91 to 8.02, or 0.89 V_{pp}. Using the AC voltage divider equation with $R_1 = 35$ ohms and $X_{C2} = 5$ ohms, AC V_{out} equals approximately 0.13 V_{pp}. Therefore, the AC variation at the output is 5.065 to 4.935 volts, a small AC ripple.

F. See Figure 11.50.

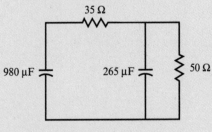

35 Ω

980 µF   265 µF   50 Ω

FIGURE 11.50

Using the simple procedure shown here always produces a working power supply circuit. This is not the only design procedure you can use for power supplies, but it is one of the simplest and most effective.

SUMMARY

This chapter introduced the following concepts and calculations related to power supplies:

- The effects of diodes on AC signals
- Methods of rectifying an AC signal
- Half-wave and full-wave rectifier circuit designs
- The calculations you can use to determine component values for half-wave and full-wave rectifier power supply circuits

SELF-TEST

These questions test your understanding of the information presented in this chapter. Use a separate sheet of paper for your diagrams or calculations. Compare your answers with the answers provided following the test.

In questions 1 through 5, draw the output waveform of each circuit. The input is given in each case.

1. See Figure 11.51.

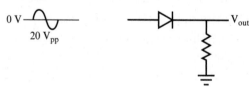

FIGURE 11.51

2. See Figure 11.52.

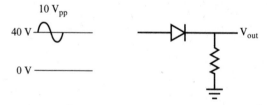

FIGURE 11.52

3. See Figure 11.53.

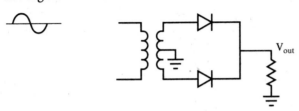

FIGURE 11.53

4. See Figure 11.54.

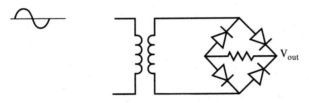

FIGURE 11.54

5. See Figure 11.55.

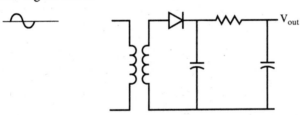

FIGURE 11.55

6. In the circuit shown in Figure 11.56, 100 V_{rms} at 60 Hz appears at the secondary coil of the transformer; 28 volts DC with as little AC ripple as possible is required across the 220-ohm load. Find R_1, C_1, and C_2. Find the approximate AC ripple.

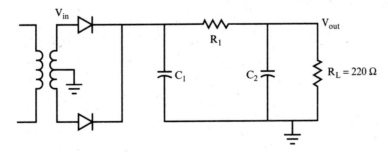

FIGURE 11.56

ANSWERS TO SELF-TEST

If your answers do not agree with those given here, review the problems indicated in parentheses.

1.	See Figure 11.57. 10 V ⌒ ⌒ ⌒ 0 V ___ **FIGURE 11.57**	(problems 1–5)
2.	See Figure 11.58. 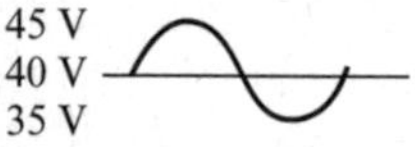 45 V 40 V 35 V **FIGURE 11.58**	(problem 2)
3.	See Figure 11.59.  **FIGURE 11.59**	(problem 11)
4.	See Figure 11.60.  **FIGURE 11.60**	(problem 13)
5.	See Figure 11.61. 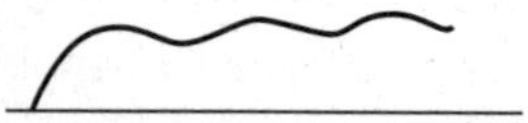 **FIGURE 11.61**	(problems 15–18)
6.	$R_1 = 833$ ohms, $C_1 = 79\ \mu F$: let $X_{C2} = 22$ ohms and then $C_2 = 60\ \mu F$ $$AC\,V_{out} = 14 \times \frac{22}{\sqrt{(22^2 \times 833^2)}} = 0.37\,V_{pp}$$	(problems 26–30)

12

Conclusion and Final Self-Test

In this book, you have discovered basic concepts and formulas that provide a foundation for your studies in modern electronics, whether you become a dedicated hobbyist or study electrical or electronics engineering.

CONCLUSION

Having read this book, you should now know enough to read intermediate-level electronics books and articles intelligently, to build electronics circuits and projects, and to pursue electronics to whatever depth and for whatever reason you want. Specifically, you should now be able to do the following:

- Recognize all the important, discrete electronics components in a schematic diagram.
- Understand how circuits that use discrete components work.
- Calculate the component values needed for circuits to function efficiently.
- Design simple circuits.
- Build simple circuits and electronics projects.

To see how much you have learned, you may want to take the final self-test at the end of this chapter. It tests your comprehension of the concepts and formulas presented throughout this book.

When you complete the following self-test and feel confident that you have mastered the information in this book, refer to Appendix E, "Supplemental Resources," for additional resources for further learning, including the following:

- Books such as *The Art of Electronics* by Paul Horowitz and Winfield Hill (New York: Cambridge University Press, 1989) provide a great next step in further electronics study.
- Magazines such as *Everyday Practical Electronics* offer interesting projects in each issue.
- You can browse websites for electronics project ideas. For example, Earl Boysen's website, `www.buildinggadgets.com`, provides tips, ideas, and links to a variety of great online resources.

NOTE   For those interested in more serious study, you should be aware that there is a difference between the path you take to become an electrician (or technician) and an electrical (or electronics) engineer. Training for electronics technicians is available in military trade schools, public and private vocational schools, and in many high schools. Engineers are required to understand the mathematical details in more depth and must take at least a 4-year curriculum at an accredited college or university.

Whatever your goal, you can feel confident that this book has given you a solid grounding for your future studies. Wherever you go in electronics, good luck!

FINAL SELF-TEST

This final test allows you to assess your overall knowledge of electronics. Answers and review references follow the test. Use a separate sheet of paper for your calculations and drawings.

1. If $R = 1\ M\Omega$ and $I = 2\ \mu A$, find the voltage. _____

2. If $V = 5$ volts and $R = 10\ k\Omega$, find the current. _____

3. If $V = 28$ volts and $I = 4$ amperes, find the resistance. _____

4. If 330 ohms and 220 ohms are connected in parallel, find the equivalent resistance. _____

5. If $V = 28$ volts and $I = 5\ mA$, find the power. _____

6. If the current through a 220-ohm resistor is 30.2 mA, what is the power dissipated by the resistor? _____

7. If the power rating of a 1000-ohm resistor is 0.5 watts, what is the maximum current that can safely flow through the resistor? _____

8. If a 10-ohm resistor is in series with a 32-ohm resistor, and the combination is across a 12-volt supply, what is the voltage drop across each resistor, and what will the two voltage drops add up to? _____

9. A current of 1 ampere splits between 6-ohm and 12-ohm resistors in parallel. Find the current through each. _____

10. A current of 273 mA splits between 330-ohm and 660-ohm resistors in parallel. Find the current through each resistor. _____

11. If R = 10 kΩ and C = 1 μF, find the time constant. _____

_____

12. If R = 1 MΩ and C = 250 μF, find the time constant. _____

_____

13. Three capacitors of 1 μF, 2 μF, and 3 μF are connected in parallel. Find the total capacitance. _____

_____

14. Three capacitors of 100 μF, 220 μF, and 220 μF are connected in series. Find the total capacitance. _____

_____

15. Three capacitors of 22 pF, 22 pF, and 33 pF are connected in series. Find the total capacitance. _____

_____

16. What is the knee voltage for a germanium diode? _____

_____

17. What is the knee voltage for a silicon diode? _____

_____

18. In the circuit shown in Figure 12.1, V_S = 5 volts and R = 1 kΩ. Find the current through the diode, I_D. _____

_____

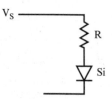

V_S

R

Si

FIGURE 12.1

19. For the circuit shown in Figure 12.1, V = 12 volts and R = 100 ohms. Find I_D. ____

_____

20. For the circuit shown in Figure 12.2, V_S = 100 volts, R_1 = 7.2 kΩ, R_2 = 4 kΩ, and V_Z = 28 volts. Find the current through the zener diode, I_Z. _____

_____

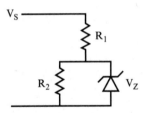

FIGURE 12.2

21. For the circuit in Figure 12.2, V_S = 10 volts, R_1 = 1 kΩ, R_2 = 10 kΩ, and V_Z = 6.3 volts. Find I_Z. _____

22. Using the circuit shown in Figure 12.3, find the DC collector voltage, V_C, if V_S = 28 volts, β = 10, R_B = 200 kΩ, and R_C = 10 kΩ. _____

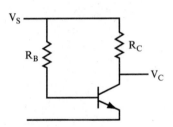

FIGURE 12.3

23. Again, using the circuit shown in Figure 12.3, find R_B if V_S = 12 volts, β = 250, R_C = 2.2 kΩ, and V_C = 6 volts. _____

24. Using the circuit shown in Figure 12.3, find β if V_S = 10 volts, R_B = 100 kΩ, R_C = 1 kΩ, and V_C = 5 volts. _____

25. What are the three terminals for a JFET called, and which one controls the operation of the JFET? _____

26. Using the circuit shown in Figure 12.4, find the value of R_B required to turn the transistor ON if V_S = 14 volts, R_C = 10 kΩ, and β = 50. _____

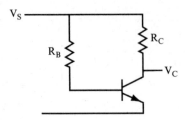

FIGURE 12.4

27. Again, using the circuit shown in Figure 12.4, find the value of R_B required to turn the transistor ON if V_S = 5 volts, R_C = 4.7 kΩ, and β = 100. _____

_____

28. Using the circuit shown in Figure 12.5, find the values of R_1, R_2, and R_3 that can enable the switch to turn Q_2 ON and OFF if V_S = 10 volts, β_1 = 50, β_2 = 20, and R_4 = 2.2 kΩ. _____

_____

_____

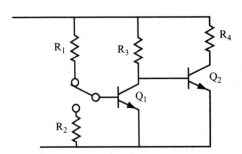

FIGURE 12.5

29. Again, using the circuit shown in Figure 12.5, find the values of R_1, R_2, and R_3 that can enable the switch to turn Q_2 ON and OFF if V_S = 28 volts, β_1 = 30, β_2 = 10, and R_4 = 220 Ω. _____

_____

30. An N-channel JFET has a drain saturation current of I_{DSS} = 14 mA. If a 28-volt drain supply is used, calculate the drain resistance, R_D. _____

_____

31. Draw one cycle of a sine wave. _____

_____

32. Mark in V_{pp}, V_{rms}, and the period of the waveform on your drawing for question 31. _

33. If V_{pp} = 10 volts, find V_{rms}. _____

_____

34. If V_{rms} = 120 volts, find V_{pp}. _____

_____

35. If the frequency of a sine wave is 14.5 kHz, what is the period of the waveform? ___

_____

36. Find the reactance X_C for a 200 µF capacitor when the frequency is 60 Hz. _____

_____

37. Find the value of the capacitance that gives a 50-ohm reactance at a frequency of 10 kHz. _____

_____

38. Find the inductive reactance X_L for a 10-mH inductor when the frequency is 440 Hz. _____

_____

39. Find the value of the inductance that has 100 ohms reactance when the frequency is 1 kHz. _____

_____

40. Find the series and parallel resonant frequency of a 0.1 µF capacitor and a 4-mH inductor that has negligible internal resistance. _____

_____

41. Using the circuit shown in Figure 12.6, find X_C, Z, V_{out}, I, tan θ, and θ, if V_{in} = 10 V_{pp}, f = 1 kHz, C = 0.1 µF, and R = 1600 ohms. _____

_____

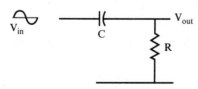

FIGURE 12.6

42. Again, using the circuit shown in Figure 12.6, find X_C, Z, V_{out}, I, tan θ, and θ, if V_{in} = 120 V_{rms}, f = 60 Hz, C = 0.33 µF, and R = 6 kΩ. _____

_____

43. Using the circuit shown in Figure 12.7, find X_C, AC V_{out}, and DC V_{out}, if V_{in} = 1 V_{pp} AC, riding on a 5-volt DC level; f = 10 kHz; R_1 = 10 kΩ; R_2 = 10 kΩ; and C = 0.2 µF. _____

_____

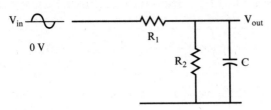

FIGURE 12.7

44. Again, using the circuit shown in Figure 12.7, find X_C, AC V_{out}, and DC V_{out}, if V_{in} = 0.5 V_{pp} AC, riding on a 10-volt DC level; f = 120 Hz; R_1 = 80 ohms; R_2 = 20 ohms; and C = 1000 µF. _____

_____

45. In the circuit shown in Figure 12.8, V_{in} = 10 V_{pp} AC, riding on a 5-volt DC level; f = 1 kHz; L = 10 mH; r = 9 ohms; and R = 54 ohms. Find AC V_{out}, DC V_{out}, X_L, Z, tan θ, and θ. _____

_____

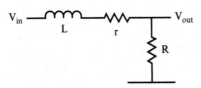

FIGURE 12.8

46. In the circuit shown in Figure 12.9, L = 1 mH, C = 0.1 µF, and R = 10 ohms. Find f_r, X_L, X_C, Z, Q, and the bandwidth. _____

_____

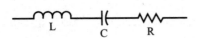

FIGURE 12.9

47. In the circuit shown in Figure 12.10, L = 10 mH, C = 0.02 µF, and r = 7 ohms. Find f_r, X_L, X_C, Z, Q, and the bandwidth. _____

_____

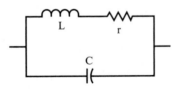

FIGURE 12.10

48. If the voltage across the resonant circuit of question 47 is at a peak value of 8 volts at the resonant frequency, what is the voltage at the half-power points and what are the half-power frequencies?_____

_____

49. Using the amplifier circuit shown in Figure 12.11, find the values of R_1, R_2, and R_E that can provide the amplifier with a voltage gain of 10. Use V_S = 28 volts, R_C = 1 kΩ, and β = 100. _____

_____

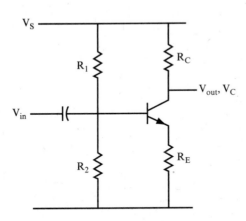

FIGURE 12.11

50. Again, using the circuit shown in Figure 12.11, find the values of R_1, R_2, and R_E that can provide the amplifier a voltage gain of 20. Use V_S = 10 volts, R_C = 2.2 kΩ, and β = 50. _____

_____

51. Using the circuit shown in Figure 12.11, how would you modify the amplifier in question 50 to obtain a maximum gain? Assume that the lowest frequency it has to pass is 50 Hz. _____

52. Using the JFET amplifier circuit shown in problem 42 of Chapter 8, "Transistor Amplifiers," with a bias point of V_{GS} = −2.8 volts, a drain current of I_D = 2.7 mA, and V_{DS} = 12 volts, find the values of R_S and R_D. _____

53. If the transconductance of the JFET used in question 52 is 4000 μmhos, what is the AC voltage gain? _____

54. A certain op-amp circuit uses an input resistance of 8 kΩ to an inverting input. For the op-amp circuit to have a gain of 85, what should the value of the feedback resistance be? _____

55. If the input to the op-amp circuit of question 54 is 2 mV, what is the output? _____

56. What is an oscillator? _____

57. Why is positive feedback rather than negative feedback necessary in an oscillator? _

58. What feedback method is used in a Colpitts oscillator? _____

59. What feedback method is used in a Hartley oscillator? _____

60. Draw the circuit of a Colpitts oscillator. _____

61. Draw the circuit of a Hartley oscillator. _____

62. What is the formula used to calculate the output frequency of an oscillator? _____

63. Draw the circuit symbol for a transformer with a center tap. _____

64. Name the two main coils used on a transformer. _____
_____

65. What is the equation that shows the relationship between the input voltage, the output voltage, and the number of turns in each coil of a transformer? _____
_____

66. What is the equation that shows the relationship between the turns ratio and the currents in the primary and secondary coils of the transformer? _____
_____

67. What is the equation that shows the relationship between the impedance of the primary coil, the impedance of the secondary coil, and the number of turns in each coil of a transformer? _____
_____

68. What are the two main uses for transformers? _____
_____

69. Draw a simple half-wave rectifier circuit with a smoothing filter at the output. ____
_____

70. Draw a simple full-wave rectifier circuit using a center tap transformer and a smoothing filter at the output. _____
_____

71. Given a 10 V_{rms} input to a full-wave rectified power supply, calculate the values of R_1, C_1 and C_2 (see Figure 12.12) that results in a 5-volt DC output across a 50-ohm load. _____
_____

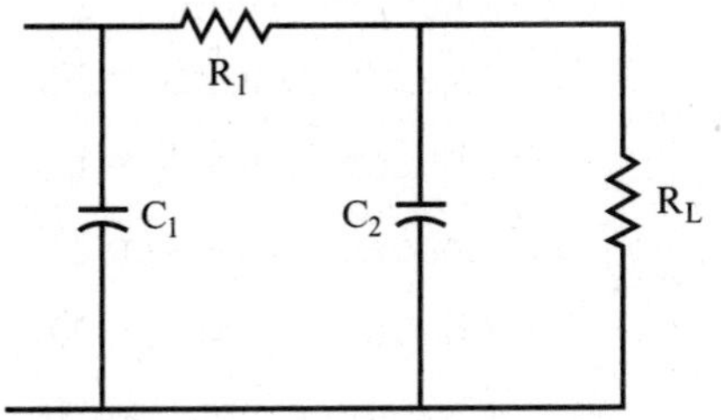

FIGURE 12.12

ANSWERS TO FINAL SELF-TEST

The references in parentheses to the right of the answers give you the chapter and problem number where the material is introduced so that you can easily review any concepts covered in the test.

1.	V = 2 volts	(Chapter 1, problem 5)
2.	I = 0.5 mA	(Chapter 1, problem 6)
3.	R = 7 ohms	(Chapter 1, problem 7)
4.	132 ohms	(Chapter 1, problem 10)
5.	P = 140 milliwatts or 0.14 watts	(Chapter 1, problems 13 and 14)
6.	0.2 watts	(Chapter 1, problems 13 and 15)
7.	22.36 mA	(Chapter 1, problems 13 and 16)
8.	2.86 volts, 9.14 volts, 12 volts	(Chapter 1, problems 23 and 26)
9.	2/3 ampere through the 6-ohm resistor; 1/3 ampere through the 12-ohm resistor	(Chapter 1, problem 28 or 29)
10.	91 mA through the 660-ohm resistor; 182 mA through the 330-ohm resistor	(Chapter 1, problem 28 or 29)
11.	τ = 0.01 seconds	(Chapter 1, problem 34)
12.	τ = 250 seconds	(Chapter 1, problem 34)
13.	6 μF	(Chapter 1, problem 40)
14.	52.4 μF	(Chapter 1, problem 41)
15.	8.25 μF	(Chapter 1, problem 41)
16.	Approximately 0.3 volts	(Chapter 2, problem 9)
17.	Approximately 0.7 volts	(Chapter 2, problem 9)
18.	I_D = 4.3 mA	(Chapter 2, problem 12)
19.	I_D = 120 mA	(Chapter 2, problem 12)
20.	I_z = 3 mA	(Chapter 2, problem 29)
21.	I_z = 3.07 mA	(Chapter 2, problem 29)
22.	V_C = 14 volts	(Chapter 3, problems 20–23)
23.	R_B = 1.1 MΩ	(Chapter 3, problems 20–23)
24.	β = 50	(Chapter 3, problems 20–23)
25.	Drain, source, and gate, with the gate acting to control the JFET	(Chapter 3, problem 28)

26.	$R_B = 500\ k\Omega$	(Chapter 4, problems 8)
27.	$R_B = 470\ k\Omega$	(Chapter 4, problems 4–8)
28.	$R_3 = 44\ k\Omega, R_1 = 2.2\ k\Omega, R_2 = 2.2\ k\Omega$	(Chapter 4, problems 19–23)
29.	$R_3 = 2.2\ k\Omega, R_1 = 66\ k\Omega, R_2 = 66\ k\Omega$	(Chapter 4, problems 19–23)
30.	$R_D = 2\ k\Omega$	(Chapter 4, problem 39)
31.	See Figure 12.13. FIGURE 12.13	(Chapter 5, problem 7)
32.	See Figure 12.14. FIGURE 12.14	(Chapter 5, problems 3 and 7)
33.	3.535 volts	(Chapter 5, problem 4)
34.	340 volts	(Chapter 5, problem 5)
35.	69 μsec	(Chapter 5, problem 7)
36.	13.3 ohms	(Chapter 5, problem 14)
37.	0.32 μF	(Chapter 5, problem 14)
38.	27.6 ohms	(Chapter 5, problem 17)
39.	16 mH	(Chapter 5, problem 17)
40.	8 kHz	(Chapter 5, problems 19 and 21)
41.	$X_C = 1.6\ k\Omega, Z = 2263$ ohms, $V_{out} = 7.07$ volts, $I = 4.4$ mA, $\tan\theta = 1, \theta = 45$ degrees	(Chapter 6, problems 10 and 23)
42.	$X_C = 8\ k\Omega, Z = 10\ k\Omega, V_{out} = 72$ volts, $I = 12$ mA, $\tan\theta = 1.33, \theta = 53.13$ degrees	(Chapter 6, problems 10 and 23)

Continued

(continued)

43.	X_C = 80 ohms, AC V_{out} = 8 mV, DC V_{out} = 2.5 volts	(Chapter 6, problem 26)
44.	X_C = 1.33 ohms, AC V_{out} = 8.3 mV, DC V_{out} = 2 volts	(Chapter 6, problem 26)
45.	X_L = 62.8 ohms, Z = 89 ohms, AC V_{out} = 6.07 volts, DC V_{out} = 4.3 volts, tan θ = 1, θ = 45 degrees	(Chapter 6, problems 31 and 35)
46.	f_r = 16 kHz, X_L = 100 ohms, X_C = 100 ohms, Z = 10 ohms, Q = 10, BW = 1.6 kHz	(Chapter 7, problems 2, 6, and 20)
47.	f_r = 11,254 Hz, X_L = X_C = 707 ohms, Z = 71.4 kΩ, Q = 101, BW = 111 Hz	(Chapter 7, problems 10, 11, and 20)
48.	V_{hp} = 5.656 volts, f_{1hp} = 11,198 Hz, f_{2hp} = 11,310 Hz	(Chapter 7, problem 27)
49.	Your values should be close to the following: R_E = 100 ohms, V_C = 14 volts, V_E = 1.4 volts, V_B = 2.1 volts, R_2 = 1.5 kΩ, R_1 = 16.8 kΩ	(Chapter 8, problem 17)
50.	R_E = 110 ohms, V_C = 5 volts, V_E = 0.25 volts, V_B = 0.95 volts, R_2 = 2.2 kΩ, R_1 = 18.1 kΩ	(Chapter 8, problem 17)
51.	The gain can be increased by using a capacitor to bypass the emitter resistor R_E; C_E = 300 μF (approximately).	(Chapter 8, problem 20)
52.	R_S = 1.04 kΩ, R_D = 3.41 kΩ	(Chapter 8, problem 42)
53.	A_v = −13.6	(Chapter 8, problem 39)
54.	R_F = 680 kΩ	(Chapter 8, problem 45)
55.	V_{out} = 170 mV and is inverted	(Chapter 8, problem 45)
56.	An oscillator is a circuit that emits a continuous sine wave output without requiring an input signal. Other types of oscillators exist that do not have sine wave outputs, but they are not discussed in this book.	(Chapter 9, introduction)
57.	Positive feedback causes the amplifier to sustain an oscillation or sine wave at the output. Negative feedback causes the amplifier to stabilize, which reduces oscillations at the output.	(Chapter 9, problems 2 and 3)

58.	A capacitive voltage divider	(Chapter 9, problem 14)
59.	An inductive voltage divider	(Chapter 9, problem 14)
60.	See Figure 12.15. 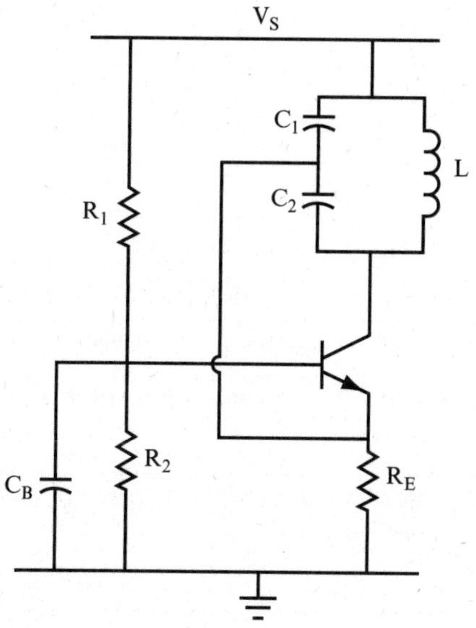 FIGURE 12.15	(Chapter 9, problem 24)
61.	See Figure 12.16. 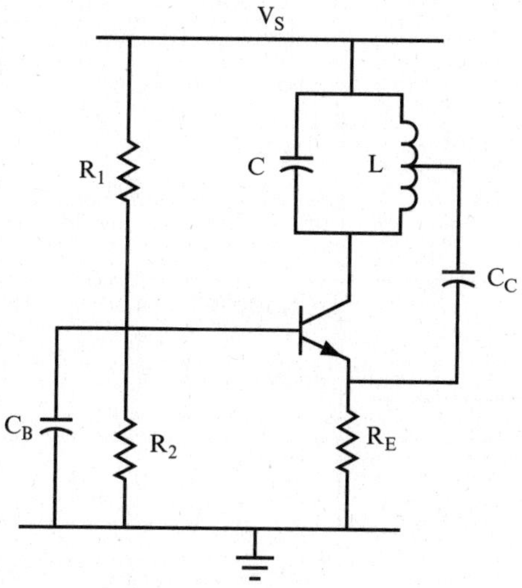 FIGURE 12.16	(Chapter 9, problem 25)
62.	$$f_r = \frac{1}{2\pi\sqrt{LC}}$$	(Chapter 9, problem 11)

Continued

(continued)

63.	See Figure 12.17.	(Chapter 10, problem 4)
	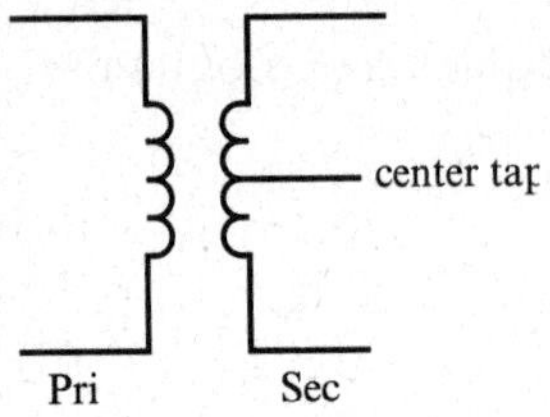 center tap Pri Sec **FIGURE 12.17**	
64.	Primary and secondary	(Chapter 10, problem 2)
65.	$V_{in}/V_{out} = V_P/V_S = N_P/N_S = TR$	(Chapter 10, problem 6)
66.	$I_{out}/I_{in} = I_S/I_P = N_P/N_S = TR$	(Chapter 10, problem 13)
67.	$Z_{in}/Z_{out} = (N_P/N_S)^2$, or impedance ratio, is the square of the turns ratio.	(Chapter 10, problem 16)
68.	They are used for stepping up or stepping down an AC voltage, and to match impedances between a generator and a load.	(Chapter 10, introduction)
69.	See Figure 12.18. 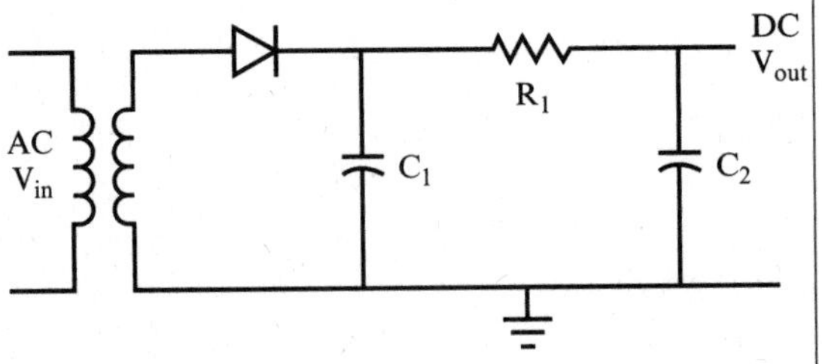 **FIGURE 12.18**	(Chapter 11, problem 14)
70.	See Figure 12.19. 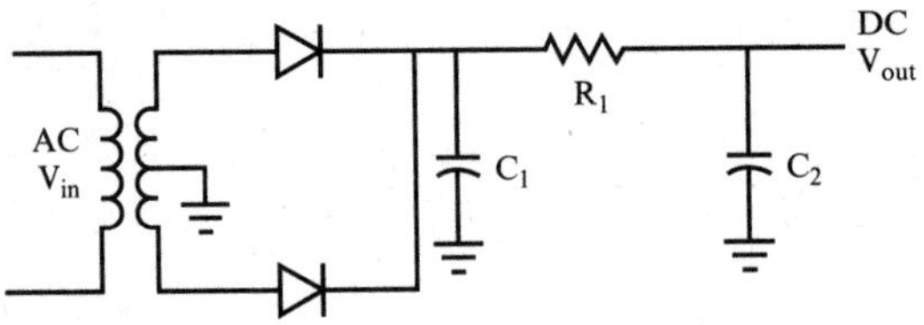 **FIGURE 12.19**	(Chapter 11, problem 31)
71.	$R_1 = 84$ ohms, $C_1 = 622\ \mu F$, $C_2 = 265\ \mu F$	(Chapter 11, problems 26–29)

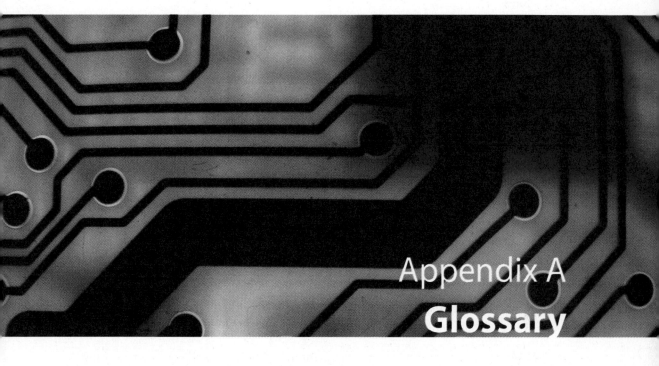

Glossary

Ampere (A)   The unit of measurement of electric current.

Amplifier   Electronic device or circuit that produces an output signal with greater power, voltage, or current than that provided by the input signal.

Capacitance (C)   The capability of a component to store an electric charge when voltage is applied across the component, measured in farads.

Capacitor   A component that stores electric charge when voltage is applied to it. It can return the charge to a circuit in the form of an electric current when the voltage is removed.

Current (I)   The amount of electric charge flowing through a circuit, measured in amperes.

Diode   A component that conducts current in one direction only.

Discrete components   Individual electronic parts such as resistors, diodes, capacitors, and transistors.

Farad (F)   The unit of measurement of capacitance.

Feedback   A connection from the output of an amplifier back to the input. In some instances, a portion of the output voltage is used to control, stabilize, or modify the operation of the amplifier. However, in some instances, unwanted feedback can cause a squealing noise or can cancel another signal.

Filter   Electronic circuit that can either block or pass frequencies that fall within a certain range.

Frequency (f)   Number of cycles of a waveform that occurs in a given time period, measured in hertz (cycles per second).

Ground   Zero volts. This is the arbitrary reference point in a circuit from which all voltage measurements are made.

Henry (H)   The unit of measurement of inductance.

Impedance (Z)   Total opposition (resistance and reactance) of a circuit to alternating current (AC) flow, measured in ohms.

Inductance (L)   The property of a component that opposes any change in an existing current, measured in henrys.

Inductor   A coil of wire whose magnetic field opposes changes in current flow when the voltage across the coil is changed.

Integrated circuit (IC)   Electronic component in the form of a small silicon chip in which numerous transistors and other components have been built to form a circuit.

Kirchhoff's laws   A set of formulas that form the basis for direct current (DC) and alternating current (AC) circuit analysis. This includes *Kirchhoff's current law (KCL)*, which states that the sum of all currents at a junction equals zero; and *Kirchhoff's voltage law (KVL)*, which states that the sum of all voltages in a loop equals zero.

Ohm (Ω)   The unit of measurement of resistance.

Ohm's law   A formula used to calculate the relationship between voltage, current, and resistance, expressed as $V = IR$. Also expressed as $E = IR$.

Operational amplifier (op-amp)   An integrated circuit, multistage amplifier. An op-amp is much smaller and, therefore, more practical than an equivalent amplifier made with discrete components.

Oscillator   An electronic circuit that produces a continuous output signal such as a sine wave or square wave.

Phase angle   For a signal, the angle of lead or lag between the current waveform and the voltage waveform, expressed in degrees.

Phase shift   The change in a phase of a signal as it passes through a circuit, such as in an amplifier.

Pinout   The configuration of leads used to connect an electronic component to a circuit.

Power   The expenditure of energy over time, measured in watts.

Reactance (X)   The degree of opposition of a component to the flow of alternating current (AC), measured in ohms. There are two types of reactance: capacitive reactance (X_C) exhibited by capacitors and inductive reactance (X_L) exhibited by inductors.

Rectification   The process to change alternating current (AC) to direct current (DC).

Resistance (R)   The degree of opposition of a component to the flow of electric current, measured in ohms.

Resistor   A component whose value is determined by the amount of opposition it has to the flow of electric current.

Resonance frequency   The frequency at which the reactance of a capacitor and inductor connected together in a circuit are equal.

Saturated transistor   A transistor that is completely turned on.

Semiconductor   A material that has electrical characteristics of a conductor or an insulator, depending on how it is treated. Silicon is the semiconductor material most commonly used in electronic components.

Transformer   A component that transforms an input alternating current (AC) voltage to either a higher level (step-up transformer) or a lower level (step-down transformer) AC voltage.

Transistor, BJT   A bipolar junction transistor (BJT) is a semiconductor component that can either be used as a switch or an amplifier. In either case, a small input signal controls the transistor, producing a much larger output signal.

Transistor, JFET A junction field effect transistor (JFET), which, like the bipolar junction transistor, can be used either as a switch or an amplifier.

Transistor, MOSFET Like the BJT and JFET, a metal oxide silicon field effect transistor (MOSFET) that can be used either as a switch or an amplifier. The MOSFET is the most commonly used transistor in integrated circuits.

Turns ratio (TR) The ratio of the number of turns in the primary or input winding of a transformer to the number of turns in the secondary (or output) winding.

Volt (V) The unit of measurement for the potential difference that causes a current to flow through a conductor.

Voltage (V) The potential difference of energy that, when applied to an electrical circuit, causes current to flow, measured in volts.

Watt (W) Unit of electric power dissipated as heat when 1 amp of current flows through a component that has 1 volt applied across it.

Zener A particular type of diode that enables the flow of current at a definite reverse-bias voltage level.

Appendix B
List of Symbols and Abbreviations

The following table lists common symbols and abbreviations.

Symbol/Abbreviation	Meaning
A	Ampere
AC	Alternating current
A_{pp}	Peak-to-peak amperes
A_{rms}	Root mean square amperes
A_V	AC voltage gain
β (beta)	Current gain
BW	Bandwidth
C	Capacitor
DC	Direct current
F	Farad
g_m	Transconductance
f	Frequency

Continued

(continued)

Symbol/Abbreviation	Meaning
f_r	Resonant frequency
H	Henry
Hz	Hertz
I	Electric current
I_B	Base current
I_C	Collector current
I_D	Drain current of a field effect transistor (FET); also current through a diode
I_{DSS}	Saturation current
I_{PP}	Peak-to-peak current
I_{rms}	Root mean square current
kHz	Kilohertz
$k\Omega$	Kilohm
kW	Kilowatt
L	Inductor
LC	Inductor-capacitor circuit
mA	Milliampere
mH	Millihenry
$M\Omega$	Megohm
μA	Microampere
μF	Microfarad
μH	Microhenry
μs	Microsecond
ms	Millisecond
mV	Millivolt
N_P	Number of turns in a primary coil
N_S	Number of turns in a secondary coil
Ω	Ohms
pF	Picofarad
P	Power
Q	Transistor; also the Q value of a resonant circuit
R	Resistor
R_{in}	Input resistance of a transistor

Symbol/Abbreviation	Meaning
r	DC resistance of an inductor
T	Period of a waveform
τ	Time constant
TR	Turns ratio
θ	Phase angle
μV	Microvolt
V	Voltage
V_C	Voltage at the collector of a transistor
V_{DD}	Drain supply voltage
V_E	Voltage at the emitter of a transistor
V_{GG}	Gate supply voltage
V_{GS}	Gate to source voltage
$V_{GS(off)}$	Gate to source cutoff voltage
V_{in}	AC voltage of an input signal
V_{out}	AC output voltage
V_p	Peak voltage
V_{PP}	Peak-to-peak voltage
V_{rms}	Root mean square voltage
V_S	Supply voltage
W	Watts
X_C	Reactance of a capacitor
X_L	Reactance of an inductor
Z	Impedance

Powers of Ten and Engineering Prefixes

The following table shows powers of the number 10, decimal equivalents, prefixes used to denote the value, symbols used, and typical usages.

Power	Decimal	Prefix	Symbol	Typical Uses
10^9	1,000,000,000	Giga-	G	GHz
10^6	1,000,000	Mega-	M	MΩ, MHz, MV
10^3	1,000	Kilo-	k	KΩ, kHz, kV
10^{-3}	0.001	Milli-	m	mA, mH, msec, mV
10^{-6}	0.000,001	Micro-	μ	μA, μF, μH, μsec, μV
10^{-9}	0.000,000,001	Nano-	n	nH, nsec
10^{-12}	0.000,000,000,001	Pico-	p	pF, pH

Standard Composition Resistor Values

The most commonly used type of resistor is the carbon film resistor with a ± 5 percent tolerance and either a 1/4 or 1/2 watt power rating. The standard resistance values for this type of resistor are listed in the following table (in ohms). You should purchase resistors at any of these values through the online distributors listed in Appendix E, "Supplemental Resources." Power resistors are available with fewer resistance values, which you can find in the catalogs of various suppliers.

NOTE In the following table, "k" represents kilo-ohms, so 7.5 k translates into 7,500 ohms. Similarly, "M" stands for megohms, so a value of 3.6 M represents 3,600,000 ohms.

2.2	24	270	3.0 k	33 k	360 k
2.4	27	300	3.3 k	36 k	390 k
2.7	30	330	3.6 k	39 k	430 k
3	33	360	3.9 k	43 k	470 k
3.3	36	390	4.3 k	47 k	510 k
3.6	39	430	4.7 k	51 k	560 k

Continued

(continued)

3.9	43	470	5.1 k	56 k	620 k
4.3	47	510	5.6 k	62 k	680 k
4.7	51	560	6.2 k	68 k	750 k
5.1	56	620	6.8 k	75 k	820 k
5.6	62	680	7.5 k	82 k	910 k
6.2	68	750	8.2 k	91 k	1.0 M
6.8	75	820	9.1 k	100 k	1.2 M
7.5	82	910	10 k	110 k	1.5 M
8.2	91	1.0 k	11 k	120 k	1.8 M
9.1	100	1.1 k	12 k	130 k	2.2 M
10	110	1.2 k	13 k	150 k	2.4 M
11	120	1.3 k	15 k	160 k	2.7 M
12	130	1.5 k	16 k	180 k	3.3 M
13	150	1.6 k	18 k	200 k	3.6 M
15	160	1.8 k	20 k	220 k	3.9 M
16	180	2.0 k	22 k	240 k	4.7 M
18	200	2.2 k	24 k	270 k	5.6 M
20	220	2.4 k	27 k	300 k	6.8 M
22	240	2.7 k	30 k	330 k	8.2 M

Appendix E
Supplemental Resources

This appendix provides a list of websites, books, magazines, tutorials, and electronics suppliers that should be of interest if you want more knowledge about basic electronics concepts, reference material for circuit design, or the supplies needed to build circuits.

WEB SITES

Following are some websites that may prove useful:

- *Building Gadgets* (www.buildinggadgets.com/)—This is an electronics reference site maintained by Earl Boysen (one of the authors of this book). There are lots of handy links to electronics tutorials, discussion forums, suppliers, and interesting electronics projects here. This website includes support web pages for each project in the *Complete Electronics Self-Teaching Guide,* including the following:

  - Data sheets for key components.

  - Parts lists with links to the appropriate locations in online catalogues for suppliers.

  - Color photos showing details of circuit construction and testing.

- *All About Circuits* (www.allaboutcircuits.com/)—This site includes an online book on electronics theory and circuits, as well as discussion forums on electronics projects, microcontrollers, and general electronics issues.

- *Williamson Labs* (www.williamson-labs.com/)—This site includes tutorials on electronics components and circuits. Many of these tutorials include animated illustrations that can help you understand how each circuit functions.

- *Electro Tech online* (www.electro-tech-online.com/)—This is a discussion forum on electronics projects and general electronics issues.

- *Electronics Lab* (www.electronics-lab.com/index.html)—This is a collection of a few hundred interesting electronics projects.

- *Discover Circuits* (www.discovercircuits.com/)—This is a collection of thousands of electronic circuits.

BOOKS

Following are some books that may prove useful:

- Electronics For Dummies, Second Edition, *by Cathleen Shamieh and Gordon McComb (Indianapolis: Wiley, 2009)*—This is a good book to start with. It provides an introduction to electronics concepts, components, circuits, and methods.

- The Art of Electronics, Second Edition, *by Paul Horowitz and Winfield Hill (New York: Cambridge University Press, 1989)*—This is a useful reference book for designing circuits, as well as understanding the functionality of existing circuits.

- ARRL Handbook for Radio Communications 2012 *(Newington, Connecticut: American Radio Relay League, 2012)*—Although this is intended for ham radio enthusiasts, it is also a useful reference book for understanding circuit design. This handbook is updated every year, but if you buy the current edition, it should be good for several years.

MAGAZINES

Following are some magazines that may prove useful:

- *Everyday Practical Electronics Magazine* (www.epemag3.com/)—This magazine provides interesting detailed electronics projects for hobbyists.

- *Nuts and Volts Magazine* (www.nutsvolts.com/)—This magazine provides information on new components for hobbyists and projects, focusing on circuits using microcontrollers.
- *EDN Magazine* (www.edn.com/)—This magazine includes articles on new components/designs for the engineering community.

SUPPLIERS

This section shows retail stores and online distributors.

RETAIL STORES

Following are some retail stores that may prove useful:

- *Radio Shack* (www.radioshack.com/)—This retail chain carries electronic components and has stores in most U.S. cities.
- *Fry's Electronics* (www.frys.com/)—This retail chain carries electronics components and has stores in nine states.

ONLINE DISTRIBUTORS

Following are some online distributors that may prove useful:

- *Jameco Electronics* (www.jameco.com/)—This is a medium-sized distributor that carries most of the components you'll need, with a reasonably sized catalog that you'll find easy to use to find components.
- *Mouser Electronics* (www.mouser.com/)—This is a large distributor that carries a wide range of components with a nice ordering system on its website that lets you put together separate orders for different projects, which is handy if you're planning multiple projects. This distributor also does a good job of packaging, clearly labeling components for shipment.
- *Digi-key* (www.digikey.com/)—This is another large distributor with a broad selection of components. Digi-key may carry components that are difficult to find at smaller suppliers.

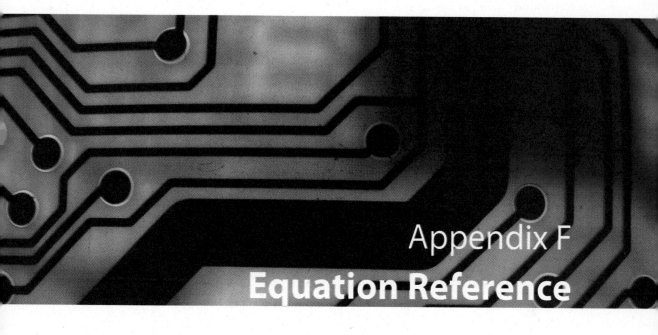

The following table provides a quick reference to common equations.

Parameter		Equation	Chapter Reference
Bandwidth		$BW = \dfrac{f_r}{Q}$	Chapter 7, problem 20
Capacitance			
	Parallel Capafitance	$C_T = C_1 + C_2 + \cdots + C_N$	Chapter 1, Summary
	Series Capacitance	$\dfrac{1}{C_T} = \dfrac{1}{C_1} + \dfrac{1}{C_2} + \cdots + \dfrac{1}{C_N}$, or $C_T = \dfrac{1}{\dfrac{1}{C_1} + \dfrac{1}{C_2} + \dfrac{1}{C_3} + \cdots + \dfrac{1}{C_N}}$, or $C_T = \dfrac{C_1 C_2}{C_1 + C_2}$ for two capacitors	Chapter 1, Summary

Continued

(continued)

Parameter		Equation	Chapter Reference
Frequency		$f = \dfrac{1}{T}$	Chapter 5, problem 7
	Resonance Frequency (series LC circuit)	$f_r = \dfrac{1}{2\pi\sqrt{LC}}$	Chapter 7, problem 6
	Resonance Frequency (parallel LC circuit)	$f_r = \dfrac{1}{2\pi\sqrt{LC}}\sqrt{1 - \dfrac{r^2C}{L}}$ if Q is less than 10, or $f_r = \dfrac{1}{2\pi\sqrt{LC}}$, if $Q \geq 10$	Chapter 7, problem 10
Gain			
	Voltage Gain	$A_v = \dfrac{V_{out}}{V_{in}}$	Chapter 8, problem 9
	Current Gain	$\beta = \dfrac{I_C}{I_B}$	Chapter 3, problem 17
Impedance		$Z = \sqrt{X_C^2 + R^2}$	Chapter 6, problem 8
Phase Shift			
	Phase Angle (RC circuit)	$\tan\theta = \dfrac{X_C}{R} = \dfrac{1}{2\pi fRC}$	Chapter 6, problem 23
	Phase Angle (LC circuit)	$\tan\theta = \dfrac{X_L}{R} = \dfrac{2\pi fL}{R}$	Chapter 6, problem 35
Q Value		$Q = \dfrac{X_L}{R}$	Chapter 7, problem 20

Parameter		Equation	Chapter Reference
Resistance			
	Parallel Resistance	$$\frac{1}{R_T} = \frac{1}{R_1} + \frac{1}{R_2} + \frac{1}{R_3} + \cdots + = \frac{1}{R_N},$$ or $$R_T = \frac{1}{\dfrac{1}{R_1} + \dfrac{1}{R_2} + \dfrac{1}{R_3} + \cdots + = \dfrac{1}{R_N}},$$ or $$RT = \frac{R_1 R_2}{R_1 + R_2}$$ for two resistors	Chapter 1, Summary
	Series Resistance	$R_T = R_1 + R_2 + \cdots + R_N$	Chapter 1, Summary
Power		$P = VI,$ or $P = I^2 R,$ or $P = \dfrac{V^2}{R}$	Chapter 1, Summary
Reactance			
	Capacitive Reactance	$X_C = \dfrac{1}{2\pi fC}$	Chapter 5, problem 13
	Inductive Reactance	$X_L = 2\pi fL$	Chapter 5, problem 16
Time Constant		$\tau = RC$	Chapter 1, Summary

Continued

(continued)

Parameter		Equation	Chapter Reference
Turns Ratio		$TR = \dfrac{N_P}{N_S}$	Chapter 10, problem 6
Voltage			
	Ohm's law (DC)	$V = IR$	Chapter 1, Summary
	Ohm's law (AC)	$V = IZ$	Chapter 6, problem 8
	Voltage divider	$V_1 = \dfrac{V_S R_1}{R_T}$	Chapter 1, Summary
	Peak-to-Peak Voltage (sine wave)	$V_{PP} = 2V_P = 2 \times \sqrt{2} \times V_{rms}$ $= 2.828 \times V_{rms}$	Chapter 5, problem 4
	RMS Voltage (sine wave)	$V_{rms} = \dfrac{1}{\sqrt{2}} \times V_P = \dfrac{1}{\sqrt{2}} \times \dfrac{V_{PP}}{2}$ $= 0.707 \times \dfrac{V_{PP}}{2}$	Chapter 5, problem 4
	Transformer output voltage	$V_{out} = \dfrac{V_{in} N_s}{N_P}$	Chapter 10, problem 7

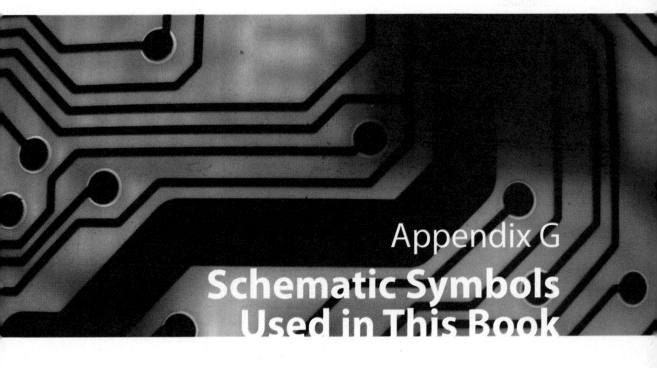

The following table shows schematic symbols used in this book.

Component		Symbol
Battery		⏚
Buzzer		⊲
Capacitors	Capacitor	⊥⊤
	Variable Capacitor	⧸

Continued

(continued)

Component		Symbol
Diodes	Diode	
	Zener Diode	
	Light Emitting Diode	
	Photodiode	
Generator (DC)		
Ground		
Inductors	Inductor	
	Tapped Inductor	
Lamp		

Component		Symbol
Meters	Meter	⌀
	Ammeter	—Ⓐ—
	Voltmeter	Ⓥ
Motor		Ⓜ
Operational Amplifier		▷
Resistors	Resistor	⌇
	Two-Contact Variable Resistor (Potentiometer)	⌇
	Three-Contact Variable Resistor (Potentiometer)	⌇←
Signal Generator (Sine Wave)		⊘
Switches	Single-pole, Single-Throw (SPST) Switch	—∘ ∘—
	Single-pole, Double-Throw (SPDT) Switch	—∘ ∘

Continued

(continued)

Component		Symbol
Test Point		
Transformers	Transformer	
	Center Tap Transformer	
Transistors	NPN BJT	
	PNP BJT	
	N-Channel JFET	
	P-Channel JFET	

DISCARD